The Biological Role of Small Leucine-Rich Proteoglycans (SLRPs) in Matrix Homeostasis and Diseases

Polly Lama
Editor

The Biological Role of Small Leucine-Rich Proteoglycans (SLRPs) in Matrix Homeostasis and Diseases

Springer

Editor
Polly Lama
Department of Anatomy and Molecular
Matrix Biology Research
Sikkim Manipal University
Gangtok, Sikkim, India

ISBN 978-981-96-7889-1 ISBN 978-981-96-7890-7 (eBook)
https://doi.org/10.1007/978-981-96-7890-7

This Springer imprint is published by the registered company Springer Nature Singapore Pte Ltd.
The registered company address is: 152 Beach Road, #21-01/04 Gateway East, Singapore 189721, Singapore

Preface

Small leucine-rich proteoglycans (SLRPs) are ubiquitous extracellular matrix molecules that play a diverse role in regulating the normal tissue homeostasis. Their expressions are either upregulated or downregulated in respone to diseases, injury, and ageing. The purpose of this book is to explain about advanced and recent analysis on the mechanisms behind synthesis, fragmentation, and dysregulation of small leucine-rich proteoglycans in common diseases of the matrix as autonomous triggers of inflammation in cancer, cartilage tissue degeneration, renal (kidney) diseases including conditions affecting the skin, tendon, musculoskeletal as well as craniofacial tissues in response to stress, injury, and biological factors that impairs normal function and development. Each chapter of this book provides a thorough explanation on SLRP family members and describes the significance behind the formations of proteolytic molecular fragments of SLRPs. It explores how these fragments alter inflammatory cell-signalling pathways to act as DAMPs (Damage associated molecular patterns), functioning similar to PAMPs (pathogen-associated molecular patterns) during early, intermediate, and advanced stages of matrix dysregulation and diseases. SLRPs and its molecular fragments interact with various cytokines in the extracellular space and bind to the cellular receptors. These interactions lead to changes in the composition and structure of SLRPs reflecting a disease state or allow initiation of matrix catabolism. Therefore, this book is valuable for undergraduate and postgraduate students, researchers, professionals, or beginners interested in investigating matrix-related changes during pathological progression of diseases with relevance to SLRPs. The book is evidence based and presents therapeutic approaches that can be useful for identification of the molecular fragments as potential biomarkers of matrix degeneration thus imparting a greater understanding on the regulatory role of SLRPs in matrix haemostasis and diseases.

This book has ten chapters, and each of them provides descriptive analytical details on the biological activity of SLRPs.

Chapter 1 introduces the family, structure, and functions of SLPRs, a distinctive group of protein present in the matrix that are characterized by a central core protein made of several leucine-rich repeats with covalently linked glycosaminoglycan side chain. There are 18 members of small leucine-rich proteoglycan protein molecules categorized in five distinct class as the SLRP family, and each of them exhibits unique or shared common features and functions essential for maintaining the integrity and stability of the molecular matrix.

Chapter 2 describes about the commonly occurring diseases of the matrix and the role of SLRPs in inflammation. SLRPs can exist in a matrix-bound, sequestered, or in a soluble form which reflects the stage of the diseased matrix. SLRPs protein core undergoes post-translational modification during tissue stress, injury, and pathological diseases breaking down the protein core into smaller fragments which are released within the matrix. These proteolytic cleaved fragments are able to interact with cellular receptors, cytokines, chemokines, growth factors, enzymes and many other matrix components in various types of diseases including progressive degenerative changes in bones, cartilages, and joints.

Chapter 3 discusses the regulatory role and significance of class I SLRPs biglycan, decorin, and asporin in extracellular matrix assembly guiding collagen fibrillogenesis, including events such as cell maturation, differentiation, and migration, thereby modulating cell signalling during inflammation-associated neo-angiogenesis and invasion of cancer cells. This chapter also explains about the immunomodulatory effect of class I SLRPs in joint disease including cardiovascular diseases and keloid disorder emphasizing their multifaceted roles in both structural matrix integrity and inflammatory regulation.

Chapter 4 presents evidence-based analytical overview of class II SLRPs fibromodulin, lumican, osteoadherin, proline/arginine leucine-rich repeat protein (PRELP), and keratocan in matrix homeostasis and catabolism. This chapter evaluates the significance of Class II SLRPs in tumour formation and in diseases of cartilage, bone, muscle, and skin indicating their therapeutic potential to act as biomarkers of disease.

Chapter 5 explains about the class III SLRPs opticin, osteoglycin/mimecan, and epiphycan, and their ability to perform certain biological function in the knockout mice model in various types of pathological conditions. Class III SLRP members show the lowest number of leucine-rich repeats, and often occur as proteolytically processed fragmented variants, and the multi-functional role of each Class III SLRPs members is discussed in this chapter.

Chapter 6 is about the non-canonical class IV SLRPs chondroadherin, nyctalopin, and tsukushi. The characteristics feature of class IV SLRPs include the lack of ear repeats. Nyctalopin is the only SLRP member with glycosylphosphatidylinositol anchored side chain attached to the C-terminus and linked to the X chromosome. Mutations in nyctalopin have been reported to cause congenital stationary night blindness (CSNB1). Chondroadherin is distinct from other members as it presents a double disulfide loop close to its C-terminus and plays a vital role in the maintenance of cartilaginous tissue. Tsukushi functions as a regulator of BMP and Wnt signalling pathways in various types of tissue. Therefore, the aim of Chap. 6 is to discuss the significance of class IV SLRPs.

Chapter 7 elucidates the characteristics of Class V SLRP with its two members, podocan and podocan-like protein-1, and although both members possess leucine-rich repeat domains, they differ in their C-terminal ends and also are considered as glycoproteins. This chapter also explores the interaction of class V SLRPs in different signalling pathways, including Wnt/β-catenin, TGF-β, and PI3K/AKT/mTOR signalling. Epigenetic regulation of podocan and podocan-like protein-1 has also been included in this chapter.

Chapter 8 ascribes the biological rationale behind molecular fragments of SLRPs and explains their role as damage-associated molecular pattern recognition entity during injury and inflammation in various tissue types. Cleaved and fragmented SLRPs represent a common end stage of pathological tissues, and these fragments can activate toll-like receptors to initiate a sterile inflammatory reaction even in the absence of blood vessels leading to upregulated expression of catabolic mediators and associated signalling pathways.

Chapter 9 demonstrates the therapeutic potential of SLRPs, emphasizing the role of their proteolytic peptides as an inhibitor or activator in the regulation of various diseases processes. For example, decorin, lumican, fibromodulin, and their fragmented peptides are shown to possess anti-tumour activity, and similarly, biglycan and its soluble peptides are involved with initiation of inflammatory cascade through a crosstalk signalling between TLR-2, 4 and purinergic P2X7 receptors, and biglycan acts as DAMPs and biomarker in various diseases of the matrix.

Chapter 10 provides a comprhensive conclusive overview of the multifunctioning roles of SLRPs emphasizing their potential function as damage associated molecular patterns (DAMPs), sensed by the innate and adaptive immune system in various tissue types. This chapter further proposes and summarizes the rationale for considering SLRPs as key inflammatory intercellular signaling molecules, highlighting their impact on immune responses and tissue homeostasis.

Gangtok, India

Polly Lama

Acknowledgements

As human beings, we all aspire to live a productive and healthy life with minimum shortcomings. However, many face health challenges due to disease such as cancer, osteoarthritis, cardiovascular disorders, and other pathologies including age-related permanent degenerative changes. A common feature behind these diseases and ageing involves alteration of the extracellular matrix molecules, where SLPRs tend to play a vital role in progression of the pathological disease. Thus, this book gratefully acknowledges the conributions of researchers, subject-magtter experts, and funding bodies particulary the Indian Council of Medical Research (ICMR/5/4-5/3/42/Neuro/2022-NCD-1) for their invaluable support in advancing our current understanding of the biological significance of small leucine-rich proteoglycans. It is hoped that readers of this book will be inspired to pursue novel, innovative, and translational approaches to detect SLRPs as molecular markers of disease, paving the way for cutting-edge diagnostics and therapeutic strategies.

Contents

Editor and Contributors

About the Editor

Polly Lama earned her PhD in molecular matrix degeneration research in cartilage diseases from the University of Bristol, United Kingdom, and subsequently served as a post-doctoral research fellow at McGill University, Canada. She is currently a Professor of anatomy, Quality assurance incharge, and Executive coordinator at Sikkim Manipal University (SMU), Sikkim, India. Dr. Lama has published several high impact papers in peer-reviewed national and international journals, including *Cell Death and Discovery*, *Nature*, *Artificial Intelligence on Medical Data*, Nature, *Bone and Joint Surgery*, and the *Journal of Anatomy*. Her contributions to research and teaching have earned her numerous awards in her field of expertise. She also serves as a reviewer and editorial member for well-known publishing bodies such as Springer Nature *BMC Musculoskeletal Diseases*, *Experimental Molecular Medicine*, *Nature*, and *Journal of Cellular and Molecular Medicine.*

Contributors

Michael A. Adams Centre for Applied Anatomy, University of Bristol, Bristol, UK

Arundhati Bag Department of Medical Biotechnology, Sikkim Manipal Institute of Medical Sciences, Sikkim Manipal University, Gangtok, Sikkim, India

Karma Lakhi Bhutia Department of Anatomy, Sikkim Manipal Institute of Medical Sciences, Sikkim Manipal University, Gangtok, Sikkim, India

Sonam Choden Bhutia Department of Biochemistry, Sikkim Manipal Institute of Medical Sciences, Sikkim Manipal University, Tadong, India

Bibhushan Dhungyal Department of Physiology, Sikkim Manipal Institute of Medical Sciences, Sikkim Manipal University, Tadong, Sikkim, India

Pramila Koirala Department of Microbiology, Nar Bahadur Bhandari Government College, Sikkim University, Tadong, Sikkim, India

Polly Lama Department of Anatomy, Sikkim Manipal Institute of Medical Sciences, Sikkim Manipal University, Gangtok, Sikkim, India

Mamta Srinivasan Medical University of Americas, Nevis, St. Kitts and Nevis, West Indies

Binod Kumar Tamang Department of Anatomy, Sikkim Manipal Institute of Medical Sciences, Sikkim Manipal University, Tadong, India

Abbreviations

A

ADAMTS	A-disintegrin and metalloproteinase with thrombospondin motifs
AKT	Serine/threonine kinase/protein kinase B
ALF	Acetaminophen-induced acute liver failure
α-SMA	Alpha-smooth muscle actin
AMPK	Adenosine monophosphate activated protein kinase
ASN	Asporin

B

β-catenin	Beta-catenin
B-CLL	B-cell chronic lymphocytic leukaemia
B4GALT-5	Beta-1,4-galactosyltransferase-5
BAD	BCL-2 associated agonist of cell death
BCL-2	B-cell leukaemia/lymphoma 2 protein
bFGF	Basic fibroblast growth factor
BGLAP	Osteocalcin
BMP	Bone morphogenetic protein
BSP	Bone sialoprotein
BYG	Biglycan

C

C1q	Complement component 1q
CAF	Cancer-associated fibroblasts
CAMKIG	Calcium/calmodulin-dependent protein kinase IG
CAV-1	Caveolin-1
CCK-8	Cell Counting Kit-8
CD	Cluster of differentiation
CGA	Cancer Genome Atlas Program

CHAD	Chondroadherin
CLL	Chronic lymphocytic leukaemia
CRC	Colorectal cancer
CRK	Cysteine-rich receptor-like kinases
CSNB	Congenital stationary night blindness
CTC	Circulating tumour cell
CTLA-4	Cytotoxic T-lymphocyte associated antigen-4
CXCL	Chemokine (cystine-amino acid-cysteine motif) ligand

D

DAMPs	Damage-associated molecular patterns
DAPC	Dystrophin associated protein complex
DCN	Decorin
DES	Diethylstilbestrol
DKK-1	Dickkopf WNT signalling pathway inhibitor-1
DMD	Duchenne muscular dystrophy
DNMT3A	DNA methyltransferase-3-alpha
DSPG3	Dermatan sulfate proteoglycan-3

E

E2IG-4	Estrogen 2 induced gene protein-4
ECM	Extracellular matrix
ECM-2	Extracellular matrix protein 2
ECM-X	Extracellular matrix protein X
EGF-R	Epidermal growth factor receptor
$E_{indsssssss}$	Effective indentation modulus
EMT	Epithelial mesenchymal transition
ENPP-2	Ectonucleotide pyrophosphatase
Epn	Epiphycan
ERK	Extracellular signal-regulated kinase
ESM-1	Endothelial specific molecule-1
EZH2	Enhancer of zeste homolog-2

F

F2RL-1	Trypsin receptor-1
FAK	Focal adhesion kinases
FAM162B	Family with Sequence Similarity 162 Member B
FMOD	Fibromodulin

G

GAGs Glycosaminoglycans
GEO Gene Expression Omnibus
GEPIA Gene Expression Profiling Interactive Analysis
GnRHa Gonadotropic-releasing hormone analogue
GO Gene ontology
GPI Glycosylphosphatidylinositol

H

HA Hyaluronic acid
HCC Hepato-cellular carcinoma
hDPSC Human dental pulp stem cell line
HIV Human immunodeficiency virus
HPV Human papilloma virus
HSPs Heat shock proteins
HTRA High temperature requirement proteases

I

ID4 DNA binding-4 tumour suppressor
IFN-γ Interferon gamma
IGF-IR Insulin-like growth factor receptor
IKKBα Inhibitor of nuclear factor kappa B kinase subunit beta
IL-1 Interleukin-1
ILL Internal limiting lamina
ISREs Interferon stimulated response elements
IVD Intervertebral disc

J

JUN Jun proto-oncogene, AP-1 transcription factor subunit

K

KEGG Kyoto Encyclopedia of Gene and Genomes
KERA Keratocan
KLF9 Kruppel-like factor-9
KSI Keratan sulphate I
KSII Keratan sulphate II

L

LCN-2	Lipocalin-2
LN	Lupus nephritis
LPS	Lipopolysaccharide
LRIT3	Leucine rich repeat Ig-like transmembrane domains-3
LRP-1	Low-density lipoprotein receptor related protein
LRRC54	Leucine rich repeat containing protein 54
LRRs	Leucine-rich repeats
LSMC	Leiomyoma smooth muscles cells
LUM	Lumican
LysC	Lysine sensitive aspartokinase-3

M

MAC	Membrane attack complex
MALDI-TOF-MS	Matrix-assisted laser desorption/ionization mass spectrometry
Met	Mesenchymal-epithelial transition factor
mGluR6	Metabotropic glutamate receptor
mi-RNA	micro-RNA
MIP-1	Macrophage inflammatory protein-1
MLC-1	Megalencephalic leukoencephalopathy subcortical cysts-1
MMPs	Matrix metalloproteinases
MSMC	Myometrial smooth muscle cells
MT1-MMP	Membrane type 1-matrix metalloproteinases
mTOR	mammalian target of rapamycin
Myc	Myelocytomatosis oncogene

N

N-WASP	Neural Wiskott-Aldrich syndrome protein
NAD	Nicotinamide adenine dinucleotide
NAFLD	Non-alcoholic fatty liver disease
NF-κB/p65	Nuclear factor kappa-B
NPY	Neuropeptide Y
NSCLC	Non-small cell lung cancer
NYX	Nyctalopin

O

OA	Osteoarthritis
OC	Ovarian cancer

OGN	Osteoglycin
OMD	Osteomodulin
OPN/SPP1	Osteopontin
OPTC	Opticin
OSAD	Osteoadherin

P

$p75^{NTR}$	p75 neurotrophin receptor
PAMPs	Pathogen associated molecular patterns
PAPS	3′-Phosphoadenosine 5′-phosphosulfate
PAX	Paired-box gene
PBMC	Peripheral blood mononuclear cells
PCAN/PODN	Podocan
PCCs	Pancreatic cancer cells
PD-1	Programmed cell death protein-1
PDAC	Pancreatic ductal adenocarcinoma
PDCD-4	Programmed cell death protein-4
PDGF	Platelet-derived growth factor
PDL	Periodontal ligament
PD-L1	Programmed cell death ligand-1
Peg3	Paternally expressed gene-3
PG-Lb	Proteoglycan-Lb
PGRN	Progranulin
PI3K	Phosphatidylinositol 3-kinase
PLAGL-2	Pleomorphic adenoma gene like-2
PLAP1	Periodontal ligament-associated protein-1
PLCG2	Phosphatidylinositol-specific phospholipase-Cγ2
PODNL-1	Podocan-like protein-1
PRAD	Prostate adenocarcinoma
PRELP	Proline/arginine-rich end LRR protein
PSCs	Pancreatic stellate cells
PSMD-2	Phosphatidylserine decarboxylase proenzyme- 2

Q

qRT-PCR	Quantitative reverse transcription polymerase chain reaction

R

RacGAP-1	Rac GTPase-activating protein-1
RANKL	Receptor activator of nuclear factor-kappa B ligand

Rap	Ras-proximate
Ras	Ras-related protein
RB-1	Retinoblastoma-1
RhoA	Ras homolog family member A
RUNX2	Runt-related transcription factor 2

S

S100A12	Extracellular receptor for advanced glycation end products
SBF-1	SET binding factor
SC 1	Stromal cell-derived matrix component-1
SCLC	Small cell lung cancer
SIRT-2	Sirtuin-2
SLL	Small lymphocytic lymphoma
SLRPs	Small leucine-rich proteoglycans
SLRR3b	Small leucine-rich repeat 3b
SLRR5A	Small leucine-rich repeat-5A
SMAD	Small mother against decapentaplegic
SMAD2	Smad-2 Mad-related protein-2
SOX	SRY-related HMG-box gene
STAT-1	Signal transduction transcriptional activator 1

T

TAA	Tumour-associated antigen
TAZ	Transcriptional coactivator with post synaptic density protein-binding motif
TCEAL5	Transcription elongation factor-A like 5
TCF/LEF	T-cell factor/lymphoid enhancer factor
TGF-β	Transforming growth factor-beta
TIA	Transient ischemic attacks
TIMPs	Tissue inhibitor of matrix metalloproteinases
TLRs	Toll-like receptors
TNFα	Tumour necrosis factor-alpha
TRAP	Tartrate-resistant acid phosphatase
TRPM-1	Transient receptor potential membrane-1
TSGs	Tumour suppressor genes
TSKU/TSK	Tsukushi

U

UUO	Unilateral ureteral obstruction

V

VAIN Vaginal intraepithelial neoplasia
VEGF Vascular endothelial growth factor
VSMCs Vascular smooth muscle cells
VSP-34 Vascular protein sorting-34
vWF von Willebrand factor

W

WISP-1 Wnt inducible secreted protein-1
WISP-1 Wnt-induced secreted protein-1
Wnt Wingless-related integration site
WST Wnt signalling inhibitor

X

X-LCSNB X-linked congenital stationary night blindness

Y

YAP Yes-associated protein

Z

ZEB-1 Zinc-finger E-box binding homeobox-1

List of Figures

The Family, Structure and Functions of Small Leucine-Rich Proteoglycans (SLRPs)

1

Polly Lama and Michael A. Adams

Abstract

Matrix organisation in a multicellular organism is essential for the maintenance of the inherent strength of structural proteins that exists between the collagens, elastins, proteoglycans, glycoproteins and the SLRPs. The intricate arrangement between the extracellular matrix molecules facilitates growth, motility and regulatory communications between the cells and its receptors. Small leucine-rich proteoglycans (SLRPs) are a distinctive type of protein present in the matrix that are characterised by a central core protein made of several leucine-rich repeats (LRRs) with covalently linked glycosaminoglycan (GAG) side chain. There are eighteen small leucine-rich proteoglycan protein molecules categorised in five distinct classes as the SLRP family, each of them exhibits a unique or common feature, and are essential for sustaining the integrity of the molecular matrix. The pericellular localisation of SLRPs allows interaction between specific cell receptors with cytokines, enzymes, growth factors, collagens, aggregating and non-aggregating large proteoglycan molecules and other vital matrix components. The diverse biological role of SLRPs makes it a viable pharmacological target and biomarker in disease progression. Thus, the main aim of this chapter is to categorically introduce each member of the SLRP family and their role in various organs and tissue-related pathological diseases associated with cell-matrix dysregulation, including the consequences of joint overloading and age-related changes in cartilaginous and non-cartilaginous tissues.

P. Lama (✉)
Department of Anatomy, Sikkim Manipal Institute of Medical Sciences, Sikkim Manipal University, Gangtok, Sikkim, India
e-mail: polly.lama@smims.smu.edu.in

M. A. Adams
Centre for Applied Anatomy, University of Bristol, Bristol, UK

P. Lama (ed.), *The Biological Role of Small Leucine-Rich Proteoglycans (SLRPs) in Matrix Homeostasis and Diseases*,
https://doi.org/10.1007/978-981-96-7890-7_1

Keywords

Small leucine rich proteoglycans · Classification · Glycosaminoglycans · Tandem repeats · Leucine · Tissues

1.1 Introduction: The Biological Importance of Small Leucine-Rich Proteoglycans (SLRPs)

Human body is a single structural entity made of billions of cells and tissues [1], called as the matrix. This organisation allows interactions and regulates maintenance of biological functions between cells and the extracellular matrix protein molecules [2]. Cell receptors and matrix components such as the SLRPs, collagens, proteoglycans and glycosaminoglycans provide an essential framework for growth, protection, maintenance of function and survival of the human body tissues [3]. Different types of cells depend on the integrity of extracellular matrix for proliferation, adherence and motility to ensure proper function and survival. Thus, isolated cells when placed in-culture in an appropriate matrix substrate attach via their matrix receptors present on the cell surface and subsequently reorganise themselves to adopt a distinctive morphology by forming focal adhesion complexes, an essential factor for cellular growth [4]. Nevertheless, during diseases, injury and ageing, the cells are exposed to changes in the matrix leading to degeneration, inflammation and dysregulation of cell signalling events.

Many of the cellular protein molecules modulating diverse matricellular functions act by forming complexes with the small leucine-rich proteoglycans (SLRPs), a unique molecule characterised by a central core protein made of several leucine-rich repeats covalently linked to the glycosaminoglycan side chain [5]. There are eighteen members in the SLRP family involved in sustaining the matrix, and they are often pericellular in spatial localisation allowing interaction with Toll-like cell receptors (TLRs) [6], insulin-like growth factor receptor (IGF-IR) [7], epidermal growth factor receptor (EGF-R) [8], low-density lipoprotein receptor-related protein (LRP-1) [9], integrin α2β1 [10, 11] and c-Met [12], acting as direct triggers for initiation of signal transduction or mechanotransduction events. These interactions sustain growth, proliferation and differentiation and allow migration of cells in developmental, physiological and pathological conditions [5, 13]. SLRPs thus form complexes with cell, their receptors, extracellular matrix molecules such as collagen, elastin, microfibrillar proteins, proteoglycans and non-collagenous glycoproteins regulating the interactions with cytokines, chemokines, enzymes, transforming growth factor beta (TGF-β), bone morphogenic protein (BMP-4), Wnt-induced secreted protein-1 (WISP-1), von Willebrand factor (VWF), platelet-derived growth factor (PDGF) and tumour necrosis factor-alpha (TNFα) [14–16]. Thus, SLRPs tend to regulate diverse biological functions in matrix homeostasis, and pharmacologic manipulation of these interactions during diseases, injury and matrix degeneration could be exploited for the treatment of proliferative, inflammatory and fibrotic disorders.

1.2 Characterisation and Classification of the SLRPs: General Features

SLRPs are categorised based on their common functional protein and gene structure [17]. They are five distinct SLRP families spanning as classes I, II, III, IV and V, Fig. 1.1. The identifying common features of SLRP protein include parameters which are (i) conservation and homology of the genes and chromosomes, (ii) protein structure and presence of characteristic N-terminal cystine-rich clusters and (iii) tandem leucine-rich repeats (LRRs) [18].

1.2.1 Conservation and Homology of SLRP Gene Structure

In humans, 18 genes encode the SLRPs that are spread over seven chromosomes. Some of these genes are clustered on the same chromosomes located between 12q21 and q23 sites, and in the comparable mouse model these clustered genes are located in a syntenic region on chromosome ten. Other non-clustered genes are mapped to the eleventh chromosomes at sites 1q32 respectively, suggesting duplication to generate functional redundancy during evolutionary changes. At the chromosomal level, the clustered and non-clustered genes may be linked together during the embryological development process, thus giving rise to overlapping of functions among SLRPs of various classes. The protein structure of SLRPs also has a variable number of tandem leucine-rich repeats (LRRs) which forms the major central domain of its protein structure [18–20].

SLRPs	CLASS I	CLASS II	CLASS III	CLASS IV	CLASS V
MEMBERS	-Biglycan (BGN) -Decorin (DCN) -Asporin (ASPN) -ECM2 -ECMX	-Fibromodulin (FMOD) -Lumican (LUM) -Osteoadherin (OMD) -PRELP -Keratocan (KERA)	-Osteoglycin (OGN) -Epiphycan (EPYC) -Opticin (OPTC)	-Tsukushi (TSK) -Chondroadherin (CHAD) -Nyctalopin (NYX)	-Podocan (PODN) -Podocan like protein-1 (PODNL-1)
SIZE (KDa)	**BGN core protein:** 42, 45 and 48 **BGN Fragments:** 39, 38, 36, 28, 27 **DCN core protein:** 43, 45, 50 **DCN Fragments:** 37, 29, 18	**FMOD core protein:** 42, 45 and 50 KDa **FMOD Fragments:** 36, 38 reported in disc **LUM core protein:** 40,45,48 Kda **LUM Fragments:** 36, 28, 27	**OGN core protein:** 43 **EPYC core protein:** 46 **OPTC core protein:** 45, 50 **OPTC Fragments:** 37kDa translocate in the nucleus of leukemia cells	**CHAD core protein:** 38, 36 **CHAD Fragments:** 35, 28 **TSK core protein:** No information found on its size	**PODN core protein:** 66
GAG SIDE CHAIN	**BGN & DCN:** 2 side chain chondroitin sulphate (CS) and dermatan sulphate (DS)	**FMOD:** Four keratan sulphate (KS) **LUM:** Four N-linked Keratan sulphate (KS)	**OGN:** Keratan sulphate **EPYC:** Chondroitin sulphate and Dermatan sulphate	**CHAD:** No side chains, CHAD provide cell attachment via α2β1 and syndecan trans membrane receptor proteins	**PODN & PODNL-1:** No side chain reported

Fig. 1.1 Small leucine-rich proteoglycans categorised into class, core protein molecular mass (KDa) and their fragmented peptide along with GAG side chains. (Figure adapted from Dr. Polly Lama, Molecular Matrix Lab)

1.2.2 Protein Structure and the N- and C-Terminal Domain

The N-terminus of SLRPs forms the starting end of a typical cluster of cysteine residue held with two disulphide bonds that stabilises the protein structure. The C-terminal is the tail sequence beyond the conserved cysteine clusters and can exist in a tissue-bound isoform. The central leucine-rich repeat domain flanked by conserved cysteine motifs on N- and C-terminus presents the LRRNT and LRRCT or LRRCE capping motif and maintains the α-helices of SLRPs as it lacks the intrahelical hydrogen bonds in the first and last turn of their protein structure. SLRPs with N-terminal capping motifs are of the LRRNT type, while C-terminal motifs can be variable. SLRPs of the canonical classes I–III show the LRRCE capping motif, while class IV SLRPs (chondroadherin and nyctalopin) have a C-terminal sequence that corresponds to LRRCT capping. Podocan, a class V SLRP, is an exception as it does not show any C-terminal capping [14, 17, 21, 22].

SLRP protein undergoes post-translational modifications involving glycosylation with glycosaminoglycan (GAG) side chains which is represented as the chondroitin sulphate, dermatan sulphate and keratan sulphate side chains, including the addition of N- or O-linked oligosaccharides and tyrosine sulphation.

[23]. Thus, the SLRP protein core tends to exhibit a typical molecular mass between 36 and 77 kDa made of tandem leucine-rich repeats with conserved leucine or asparagine residue, glycosaminoglycans side chains and alternating hydrophobic and hydrophilic amino acids. The similarities in sequence homologies of SLRPs are most pronounced in the central domain of its protein structure [24].

1.2.3 Tandem Repeats

They are repetitive protein sequence, rich in leucine protein in the SLRP molecules, while the other small hydrophobic residues are represented by various amino acids with sequence pattern as LxxLxLxxNxL, where 'x' can be any other amino acid along with the conserved leucine (L) molecule, occasionally substituted by isoleucine, valine or less frequently by hydrophobic amino acids. Lengths of the leucine-rich repeats in each SLRP class may vary from 20 to 39 amino acids. LRRs of classes I, II and V SLRP show a typical model following a short–long–long protein chain pattern throughout the entire LRR domain [25].

The other typical feature of class I–III SLRPs is the presence of the ear repeat, and class IV and V members do not present this type of modification [26]. The ear repeat is also considered as the hallmark feature of the true SLRPs based on its structural organisation. The LRRCE capping motif surrounds the ear repeat, while the final β-sheet closes the helical domain; however, the significance of ear repeats can differ at a functional level. For example, tsukushi, a class IV SLRP without the ear repeats, is a potent modulator of bone morphogenetic protein (BMP) and overlaps the functional role as shown by biglycan, a class I SLRP molecule [27, 28]. Similarly, podocan, a class V SLRP lacking ear repeats, binds with collagen I and inhibits cell growth which is similar to the biological activity of class I–III SLRPs [21].

Leucine-rich repeat (LRR) binding of SLRPs is found to be intracellular, at a cell surface level, and even as an extracellular protein domain. Crystal structures of the LRR binding domains show they adopt a curved solenoid fold, with parallel β-sheet forming the inner concave surface, while the outer convex surface forms a variety of secondary structural topologies or dimers. The LRR domains on the concave surface can contain the protein ligand-binding sites giving rise to diversity in its functions [21, 29].

These similarities in structure between different classes of SLRPs suggest they share common biological functions. Conversely, differences in their regulatory expression pattern indicate SLRP functions are modulated in a cell- or tissue-specific manner. Thus, SLRPs comprise of eighteen members classified into five different classes named as I, II, III, IV and V, Fig. 1.1, and each of them is described in detail as categories from class I to V.

1.2.4 Class I SLRPs

Class I SLRPs are decorin (DCN), biglycan (BYG), asporin (ASN), extracellular matrix protein 2 (ECM-2) and extracellular matrix protein X (ECM-X). A common feature of the class I SLRPs is the similar exonic organisation with the highly conserved intron/exon junctions, and the central domain of class I SLRP shows the characteristic presence of N-terminal cysteine cluster that forms two disulphide bonds [17].

1.2.4.1 Decorin (DCN)

Decorin also designated as proteoglycan 2 or proteoglycan 40 SLRP is present in various types of connective tissues and organs. The decorin gene is approximately 45 kbs, Fig. 1.1, located on chromosome 12ql3.2 and has eight exon gene structures. The term 'decorin' is derived from its ability to bind to collagen fibres where it 'decorates' the collagen surfaces. The core protein size is approximately between 36 kDa or 38 kDa attached to the N-linked and/or O-linked oligosaccharides with the dermatan sulphate glycosaminoglycan side chain of 80 kDa at sernine-4 amino acid site. The presence of large glycosaminoglycan side chain to the decorin core protein facilitates binding of a specific surface collagen fibril. The presence of the dermatan sulphate side chain also allows it to interact with another dermatan sulphate chain from the adjacent decorin molecule that ultimatley binds to the collagen fibril and forms a three-dimensional framework. Interactions between the collagen, decorin and dermatan sulphate side chain impart adherence between collagen fibrils that regulate the relative spacing between these molecular complexes. Thus, decorin is ubiquitously expressed by connective tissues, and alterations in its structure affect the integrity of tissues such as skin, bones, cartilages, kidney, urinary bladder including many other organs and tissue types. The importance of decorin together with biglycan (another class I SLRP) in the formation of connective tissues is demonstrated by the anomalies observed in mice that fail to express these genes. Thus, decorin (D)-deficient mice have fragile skin attributable to coarse and irregular collagen fibril spacing, confirming their importance in normal collagen fibrillogenesis [30–32].

1.2.4.2 Biglycan (BGN)

Biglycan is the second type of SLRP class I molecule also termed as proteoglycan 1 or leucine-rich proteoglycan 1 molecules. The biglycan gene is approximately 8 kbs located on chromosome Xq28 and like the decorin has an eight-exon gene structure. Spatially it is localised around the cell surface as a pericellular matrix proteoglycan and is similar in structure to the decorin (class I) and fibromodulin (class II) SLRP molecules. The core protein size is approximately 39–42 kDa, Fig. 1.1, made up of a series of eleven tandem leucine-rich repeats of 24-amino-acid sequence, containing hydrophobic residues at conserved locations. Biglycan contains two glycosaminoglycan side chains, thus named as 'bi' (two) 'glycan' (side chain). These side chains are attached to the amino acids 5 and 10 of the secreted protein structure [33, 34]. The relationship of biglycan to a number of other proteins containing the leucine-rich repeats is due to the homologies of cysteine region immediately adjacent to the leucine repeat sequences; thus, biglycan throughout evolution has been shown to exhibit protein–protein, protein–cell or cell–cell interactions. Biglycan also has been reported to bind with transforming growth factor beta in vitro and is localised on the cell surface or the pericellular spaces allowing it to interact with a variety of connective tissue cell types including epidermal cells, endothelial blood vessels cells, kidney collecting tubular cells, skeletal myofibrils, bone cells and cartilage chondrocyte cells. Biglycan, unlike decorin, is not directly associated with the regulation of collagen fibrils [35, 36].

Biglycan-deficient mice show reduced longitudinal bone growth and decreased bone mass, indicating an important role of biglycan in maintaining bone health. Mice lacking decorin and biglycan genes exhibit profound osteopenia compared to mice deficient in only one of these SLRPs [37, 38].

Biglycan gene consists of eight exons including one in the sequence that encodes the 5-untranslated region of the mRNA. The first and seventh introns are approximately 1 kilobase pair, while the remainder are shorter [35]. With the exception of the first two introns, all of the introns are spread throughout the hydrophobic repeat domain. The 500-base pair 5′ starts the transcription of biglycan and their LRR sequence, which suggest that it contains a significant amount of the gene promoter. The biglycan promoter gene also lacks the conserved CAAT and TATA box DNA sequence which are present in the promoter region, but biglycan still presents a rich glycosaminoglycan content rich side chains.

1.2.4.3 Asporin (ASPN)

Asporin also called as periodontal ligament-associated protein-1 (PLAP1) is encoded by a gene present at the human chromosome site 9q22-9q21.3. Unique to class I SLRPs, asporin is named due to the presence of aspartic acid repeats at the N-terminus of its protein structure. Asporin lacks the typical serine–glycine dipeptide and the side chains required for glycation. The structure of asporin has resemblance through its N-terminus with class II (osteoadherin), class III (epiphycan) and class V (podocan) members; thus, ASPN is also referred to as 'not a true proteoglycan'. Asporin much like the decorin molecule binds to type I collagen fibres, and they are suggested to compete for the same collagen binding sites. Asporin also binds to transforming growth factor beta-1 (TGF-β1) and calcium signalling genes in vivo; thus, overexpression of asporin gene especially during repeated loading decreases TGF-β1-mediated chondrogenesis in cartilage tissue that influences calcium uptake in mineralising tissue. The asporin protein gene is uniquely present in cluster and even contains the extracellular matrix protein-2 gene, osteoadherin (class II SLRP) and osteoglycin/mimecan (class III SLRP) genes [39–41].

Asporin protein has been found to be expressed in the human and mouse model in the perichondrium or periosteum of maxilla, vertebrae, mandible and long bones but is absent in the epiphyseal cartilage. In humans, it is also reported to be expressed in aorta, uterus, heart and liver and are particularly abundant in articular cartilage, periodontal ligament (PDL) and the dental follicle [39, 42]. A proteomic analytical study reports that asporin content in knee meniscus is much higher than in the medial tibial condyle and the cartilage of the femoral head [43]. Like decorin and fibromodulin, asporin directly binds to TGF-β1 and collagen fibrils through the LRR domain [21, 44, 45]. The N-terminus also interacts with many other matrix constituents including collagen fibres, thus playing a role in collagen fibrillogenesis and metabolism [18].

1.2.4.4 Extracellular Matrix Protein-2 (ECM-2)

Extracellular matrix protein-2 also called as stromal cell-derived matrix component (SC 1) though much larger and structurally different from conventional SLRPs shows the characteristic leucine-rich repeats (LRRs) which are 45% identical to the decorin molecule. ECM-2 is also considered to be a paralog of lumican gene physically linked to asporin on the chromosome site 9q22-9q21.3. ECM-2 presents a heparin and collagen-binding domain, and the deduced amino acid sequence of ECM-2 suggests that it is a secreted calcium-binding protein and a matrix component of the brain tissue [18, 40, 46]. A human homolog of ECM-2 has been isolated as a monoclonal antibody that shows spatial association with basal, lateral and apical surfaces of endothelial venule, and is considered to be anti-adhesive on fibronectin-coated surfaces [47].

1.2.5 Class II SLRP

Class II SLRP molecule includes fibromodulin, lumican, proline/arginine-rich end LRR protein, keratocan and osteoadherin. Based on the gene homology, class II SLRP protein molecules have a similar three-exon organisation with a large central domain that codes most of the LRRs [17, 36]. A distinct post-translational modification of class II members is characterised by the addition of sulfotyrosine residues at their N-terminus contributing towards their polyanionic properties. A sulphonate group from the donor 3′-phosphoadenosine 5′-phosphosulfate (PAPS) is transferred to the hydroxyl group as a peptidyl tyrosine residue. This sulphation reaction is catalysed by tyrosylprotein sulfotransferase 1 or 2, and the sulfotyrosine (sTyr, sY) residue is generally known to mediate protein–protein interactions critical for maintaining a diverse range of physiological functions [3]. Class II members also contain keratan sulphate side chain and polylactosamine, an unsulphated form of keratan sulphate chain. Except for proline/arginine-rich end leucine-rich repeat protein (PRELP), all class II and even III SLRPs show tyrosine sulphations, and in case of fibromodulin and lumican this sulphation site is found in the N-terminus region [21]. Such variabilities in the terminal amino acid sequence provide specific functions to class II SLRP members through different structural attachments ranging as the clusters of tyrosines, acidic, basic amino acid residues, GAG chains and other polysaccharides.

1.2.5.1 Fibromodulin (FMOD)

Fibromodulin also named as keratan sulphate proteoglycan (KSPG) is encoded by a protein gene present at the human chromosome site 1q32.1 with four exons and is approximately 8.5 kbs. It possesses a central leucine-rich repeat region with distinct four keratan sulphate side chains flanked by terminal domain containing the disulphide bonds. Owing to the interaction with type I and type II collagen fibrils, the encoded protein is suggested to play a role in the assembly of the extracellular matrix [48, 49]. FMOD also regulates TGF-beta activities by sequestering it, and variations in its gene structure have been associated with the pathogenesis such as high myopia [50]. Alternative splicing of the gene exons results in multiple transcript variants, suggesting FMOD can exist as a cytosolic and secretory collagen-binding protein characterised by a core protein size of approximately 45–59 kDa [51]. The protein core is substituted covalently with the keratan sulphate glycosaminoglycan (GAG) side chains, Fig. 1.1, and was first reported in the articular cartilages [52, 53]. Recent t studies show its expression during foetal skin development and scarless tissue repair [54]; thus, fibromodulin is widely expressed in connective tissues and plays a central role in the organisation of collagen fibril structure. Due to its abundance in the extracellular matrix of specialised connective tissue such as cartilages, ligaments and tendons, the significance of fibromodulin during wound repair associates with underlying condition of the cells and the corresponding matrix tissue [55].

1.2.5.2 Lumican (LUM)

Lumican gene resides on chromosome 12q22 and significantly overlaps with the fibromodulin gene structure composing of three exons with the first intron preceding the translation initiation site, while the second intron antecedes the termination codon [56]. The lumican protein core has a size of 38–45 kDa with a negatively charged N-terminus region containing tyrosine sulphate with disulphide bonds, six to ten LRR motifs as the common leucine-rich repeats, and a C-terminus region with two conserved cysteine residues [20]. Amino acid sequencing shows the presence of four possible keratan sulphate side chains or oligosaccharides within the leucine-rich repeat region, Fig. 1.1. In the juvenile cartilages, lumican exists as a keratan sulphate proteoglycan with sulphated polylactosamine chains [57]. However, as age advances, lumican can exist as a glycoprotein, devoid of polylactosamine chains [58]. The glycoprotein forms of lumican are present during early embryonic development of cornea, prior to its substitution with the keratan sulphate side chain. The predominance of the glycoprotein form of lumican in adult cartilage suggests it is not the product of glycosidase action since the keratan sulphate side chains of the aggrecan protein molecules are much longer in the adult than in the juvenile. Therefore, it has been proposed that with age the human articular cartilage chondrocyte cells switch the synthesis of the glycosylated proteoglycan form of lumican to a non-glycosylated version leading to decrease in the synthesis of keratan sulphate side chains [59].

Apart from the cartilages, lumican is the major keratan sulphate proteoglycan of the cornea, and regulates collagen fibril organisation, circumferential bone growth, corneal transparency, epithelial cell migration and even tissue repair [60]. Lumican has also been shown to be expressed in organs such as the gall bladder, urinary bladder, adrenal gland, appendix, endometrium layer of uterus, ovaries, dermis of the skin, heart, liver and placenta indicating a wide distribution of this SLRP [61–64]. Thus, lumican knockout mice are characterised by an abnormal connective tissue phenotype with fragile skin and cloudy cornea with thick or irregular collagen fibrils. Recently, other functions of lumican have been demonstrated involving regulation of angiogenesis, cell-cycle initiation, cell proliferation, cell migration, Toll-like receptor signalling and Fas-ligand meditated immune cell activation indicating its significance in the maintenance of matrix functions [65, 66].

1.2.5.3 Proline/Arginine-Rich End LRR Protein (PRELP)

Proline/arginine-rich end LRR protein also called as prolargin or proline/arginine-rich end leucine-rich repeat protein (SLRR2A) is structurally similar to fibromodulin and the gene encoding it located about 80 kbs proximal to FMOD at the chromosome site 1q32.1 [67]. PRELP is a secreted glycoprotein synthesised as a 20-amino-acid protein sequence containing an N-terminus extension with the leucine-rich repeats (LRRs) and an additional proline amino acid imparting (s) conformational structural rigidity. The N-terminus with the LRRs also binds negatively to the heparin/heparin sulphate, chondroitin sulphate and gram-negative bacterial cell walls indicating its participation in protein–protein interactions. PRELP has a core protein size of 55–58 kDa, which was primarily shown to be synthesised by bone osteoblasts, skeletal muscle and chondrocytes [68]. Recent evidence also confirms its presence in the basement membrane of the kidney Bowman's capsule, epididymal epithelium of male reproductive organs and the stratified squamous epithelium of the skin [69]. In cartilages, PRELP links chondrocyte cell membrane with heparin sulphate side chains and the endogenous type II collagen fibers, while in the basement membrane, PRELP anchors with type IV collagen and laminin, further linked together by nidogen reinforcing its structure [69, 70]. PRELP in the basement membrane also binds to free heparan sulphate side chains via the N-terminus and underlying collagen fibres by the LRR repeats. The bone osteoblast cell-secreted PRELP undergoes proteolysis by enzymes such as lysine-sensitive aspartokinase-3 (LysC) and glutamyl endopeptidase generating a distinct 40–75 kDa N-terminal fragments that can bind to chondroitin sulphate side chain, especially on the precursor osteoclast cell surfaces [71]. The smaller PRELP fragment is internalised inside the cells, and forms complexes with annexin-II, which is further translocated to the cell nucleus where it interacts with nuclear factor kappa-B [72] to block osteoclast maturation [72, 73]. PRELP in connective tissues form complexes with aggrecan globular-3 domain through the c-reactive protein repeat and in the synovial fluid it is shown to be involved with the regulation of the complement immune system [73]. Thus, PRELP undergoes proteolytic processing in the bone during inflammation in an animal model to release an N-terminal fragment of size 21–42 kDa [69, 70]. This sequence has been shown to possess potent antimicrobial activity in the bacterial cell walls. Mature human PRELP shares 91% similarities with the mouse PRELP structure [74].

1.2.5.4 Keratocan (KERA)

Keratocan also called as keratan sulphate proteoglycan is characteristically expressed in the cornea. The sequence of human keratocan (KERA) gene is spread over 7.65 kbs of DNA at chromosome 12q22 site with two introns and three exons. Exon 1 remains untranslated, while exon 2 contains the start codon, followed by the N-terminus peptide with a highly conserved region containing ten leucine-rich repeats (LRRs). Exon 3 forms the precursor keratocan protein, while the introns remain as a non-coding sequence [75]. The presence of KERA gene in human cornea has been linked to developing and maintaining corneal transparency as it regulates collagen fibrillogenesis. Mutations in the KERA gene are a cause behind autosomal cornea plana type 2 defect characterised by a flattened corneal surface leading to a decrease in refraction, reduced visual activity, strong hyperopia, widened limbus zone, opacities in the corneal parenchyma and marked arcus senilis [75]. Keratocan is also expressed, at lower levels, in the skin, ligaments, tendons, cartilage, arteries and striated muscles [76]. The bone osteoblast cells have also been shown to secrete keratocan to modulate osteogenic differentiation [77]. A peculiar feature of keratocan is the presence of keratan sulphate side chains which are of two types named as keratan sulphate I (KSI) and keratan sulphate II (KSII) glycosaminoglycans [78], and based on their structural differences, the corneal keratan sulphate chain is linked to the presence of asparagine in the N-terminus, whereas in cartilages KSII is O-linked to serine threonine residues [79].

1.2.5.5 Osteoadherin (OSAD)

Osteoadherin is also termed as osteomodulin (OMD) or small leucine-rich repeat 2C (SLRR2C). The human osteoadherin gene is located at the chromosome 9q22.31 site with three exons. It encodes a 421-amino-acid precursor that contains a 20-amino-acid protein sequence and eleven tandem leucine-rich repeats with keratan sulphate proteoglycan as the side chain [80, 81]. Structure-wise the leucine-rich repeat (LRR) motifs show a conserved leucine spacing folded into one beta-sheet and one alpha-helix. OSAD also contains eight sulphated tyrosine residues and is distinguished from other class II SLRPs by the presence of an approximately 70-amino-acid C-terminus acidic domain rich in aspartic acid (asparagine) and glutamic acid (glutamine). The acidic terminal of OSAD imparts tight binding to the hydroxyapatite mineral molecule present in the bone and is predominantly expressed by foetal and adult bone osteoblast cells but they are undetectable in cartilage or tendons [82]. OSAD promotes the adhesion of bone cells osteoblasts and odontoblasts to the surrounding matrix. In dentine tissue, osteoadherin is expressed by odontoblast and ameloblast cells involved in the mineralisation of teeth [83]. Human OSAD shares 80–84% amino acid sequence identity with bovine, mouse and rat OSAD [39] and 32–35% amino acid sequence identity with other human class II SLRPs such as the fibromodulin, keratocan, lumican and PRELP. OSAD protein core is expressed as a large molecule of 110 kDa in dentine tissues, and in bone osteoadherin protein core has a size of 47 kDa that promotes adhesion of osteoblasts and odontoblasts to the surrounding matrix mediated by integrin alphaV beta 3 [84]. The expression of osteoadherin gene in the bone is also associated with osteoblast cell differentiation and is suggested to play a role in the apoptosis and growth of these cells [82].

1.2.6 Class III SLRP

Class III SLRP members are opticin, osteoglycin/mimecan and epiphycan, and they show the lowest number of leucine-rich repeats, and often occur as proteolytically processed fragmented variants. Recent evidence also indicates SLRP III members can compensate mutually to perform certain biological function in the knockout mice model regulating essential process such as collagen fibrillogenesis [17]. The role of each class III SLRP members is discussed in the following subsections.

1.2.6.1 Opticin (OPTC)

Opticin also called as ‘oculoglycan’ is encoded by the OPTC gene localised to the chromosome site 1q31-q32 [85]. OPTC is a glycosylated protein with a sialylated O-linked oligosaccharide covalently attached to the protein core at the reactive amino/N-terminus. Opticin are secreted into the extracellular matrix of various tissues but are predominantly associated with the ocular (eye) tissues specially in the cornea, iris, ciliary body, vitreous humour, choroid and the retina region of the eyes [86]. Opticin was first extracted from the surface of bovine vitreous humour containing collagen fibrils, and recent evidence indicates that it binds to various types of growth hormone to modulate its functions [86, 87]. As a class III SLRP member, it is characterised by the unique seven tandem leucine-rich repeats (as opposed to 12) flanked by cysteine clusters without any glycosaminoglycan side chain; therefore, opticin is also considered as a glycoprotein with an amino terminus region showing a cluster of sialylated O-linked oligosaccharides [18].

Owing to its unique structural configuration, opticin is considered an exceptional member of the SLRP family but shares common function associated with collagen fibril assembly and spacing, including role in cell growth and proliferation [88]. In the vitreous humour of the eye, extracellular OPTC protein has also been shown to be anti-angiogenic as it binds to collagen and inhibits integrin-mediated α2β1 and α1β1 endothelial cell adhesion [89]. The molecular weight of the normal glycosylated opticin core protein varies approximately between 45 kDa and 62 kDa that undergoes degradation into smaller sizes by proteases such as matrix metalloproteinases (MMPs) 1, 2, 3, 7, 8, 9 and 13, ADAMTS4 and ADAMTS5, Fig. 1.1, contributing to the pathogenesis in ocular diseases and osteoarthritis [85, 90]. Newer investigations also indicates the presence of OPTC in non-ocular tissues such as the brain, skin, ligament and cartilage [87]. In normal ocular tissues, opticin is suggested to act as a biological glue between the vitreous collagen fibrils and the internal limiting lamina (ILL) of retina and the eye lens; therefore, the role of opticin as a therapeutic target for the pharmacological intervention in posterior vitreous detachment of the eyes has been suggested [91]. Further, mutations in the opticin gene are associated with age-related macular degeneration, one of the commonest causes of blindness in the world [85]. Opticin also forms dimers with decorin class I SLRP, and this unique three-dimensional stoichiometric ratio allows bivalent interaction between them to enable affinity for ligand binding, including interactions with substrates, inhibitors, activators, signalling molecules and neurotransmitters [92].

1.2.6.2 Osteoglycin (OGN)

Osteoglycin is also called as mimecan or osteoinductive factor has a single-copy human gene located at chromosome site 9q22.31 [93]. It was first isolated as a protein that was found to induce ectopic bone formation when inoculated into rat skeletal muscle [94]. OGN produces three mRNA transcripts as a result of differential splicing inside the first translated exon [95]. Osteoglycin has a unique cysteine-rich sequence with only six to ten leucine-rich repeats (LRRs), and the N-terminus bears keratan sulphate side chains that are attached to the LRR domains with numerous glycosylation sites [17]. OGN has a protein core size between 34 kDa and 40 kDa, Fig. 1.1, and it has been extracted from tissues such as the bone, cornea, skin, aorta, sclera, vagus nerve and cartilage and at a lower level from the ovary, cerebellum, skeletal muscle, heart and kidney tissue [96]. OGN is secreted into the extracellular matrix after post-translational modification; thus, it interacts with fibrillar collagens, growth factors, various matrix molecules and cell membrane receptors. In the bone, OGN regulates osteoblastic cell differentiation and collagen fibrillogenesis and is highly expressed in cardiovascular diseases, connective tissue disorder and cancer [97]. Osteoglycin protein is shown to be associated with ectopic bone formations, and in vivo studies suggest bone cell receptor is activated with transforming growth factor-β and bone morphogenetic proteins 2 and 3 [98], and the interaction between OGN and TGF-β pathways accelerates the synthesis of anabolic factors regulating bone formations. TGF-β produced by osteoblasts regulates and maintains bone metabolism through type I membrane receptor that promotes the formation of dimers with the type II receptor stimulating serine/threonine kinase activity causing phosphorylation of Smad2 and Smad3; this process is considered to be essential for osteoblastic bone formation.

Osteoglycin also acts as a component of the vascular matrix, produced by the vascular smooth muscle cells, and they are downregulated by pro-angiogenic factors [99]. In colorectal cancer, OGN acts as a tumour suppressor by enhancing T-lymphocyte recruitment with decreased expression of pro-angiogenic factors such as vascular endothelial growth factor (VEGF) that represses proinflammatory upregulation of chemokines and activation of endothelial cell enhancing T-lymphocyte functions [100]. The ability of osteoglycin to impede angiogenesis while promoting immune system activation for recruitment of T-lymphocyte suggests that it is a potent onco-suppressive molecule in colorectal cancer that can be explored for further therapeutic investigation in controlled conditions where osteoglycin does not bind to other signalling receptors.

1.2.6.3 Epiphycan (EPN)

Epiphycan is also called as dermatan sulphate proteoglycan-3 (DSPG3), proteoglycan-Lb (PG-Lb) or small leucine-rich repeat 3b (SLRR3b). EPN gene is located at the human chromosome 12q21.33 site and is composed of seven exons [101]. Epiphycan is a secreted SLRP molecule with a core protein size of 46 kDa, attached to the N-terminus by the dermatan sulphate side chain and three O-linked oligosaccharides at serine or threonine residues in the translated cDNA sequence [102]. The size of the glycosaminoglycan side chain additionally varies from 23 kDa to 34 kDa, with a four-cysteine cluster present at the N-terminus of the leucine-rich repeats (LRRs), along with a disulphide-bonded cysteine pair present at the C-terminus of the protein core. Epiphycan has seven to ten leucine-rich repeats of heterogeneous length that exists between the two C-terminals.

Epiphycan is expressed in cartilage, ligament, placenta, testis, cornea and the epiphyseal plate of long bone [103]. It plays a role in bone formation and in maintaining the ordered structure of cartilage tissue through matrix organisation. EPN regulates collagen fibrillogenesis by interacting with various other collagens and the extracellular matrix proteins [104]. Epiphycan is also shown to function in the regions of cartilage that is not associated with calcification where it acts to delay the onset of calcification or arranges the matrix for extensive remodelling occuring during the calcification process; thus, type II collagen is shown to be completely removed or replaced with a calcified type I collagen fibers [43, 102].

In chick and mouse cartilage tissue, EPN is involved with chondrocyte differentiation and cartilage stability. Epiphycan-deficient mice demonstrate it is critical for growth of long bones and maintenance of joint functions influencing femur length in male mice but is not essential for embryonic skeletal development. Furthermore, EPN is necessary for retaining cartilage integrity with age because the loss of epiphycan and absence of biglycan (class I SLRP) significantly accelerate the rate of osteoarthritis like lesion development when compared to single SLRP-deficient mice. The rapid progression of osteoarthritis in the double-deficient mouse suggests two possibilities: (1) Epiphycan and biglycan have some overlapping functions within cartilage. (2) Epiphycan and biglycan have independent functions and, when deleted, exacerbate the progression of osteoarthritis. Further research would be necessary to demonstrate which possibility is correct in human tissue, and present studies on in vivo gene interaction suggest involvement of both class I and class III SLRPs in osteoarthritis [104].

Epiphycan gene along with keratocan, lumican and decorin SLRPs is associated with posterior amorphous corneal dystrophy, a rare, autosomal dominant disorder that typically demonstrates partial or complete posterior lamellar corneal opacification, decreased corneal thickness and flattening of the corneal curvature [105]. The pathways related with the diseases are suggested to involve integrin receptors and ERK signalling mechanism where EPN is essential for the formation of transparent cornea necessary for imparting vision.

1.2.7 Class IV SLRPs

Class IV SLRPs comprise of chondroadherin, tsukushi and nyctalopin, they are non-canonical and structurally diverse members that share a common evolutionary origin [5]. The characteristic feature shows the absence of the ear repeats [17], and they present a protein core structure with eleven leucine-rich repeats attached to the N-terminus cysteine-rich region. Nyctalopin is the only SLRP member with glycosylphosphatidylinositol (GPI)-anchored side chain attached to the C-terminus and linked to the X chromosomes [106]. Mutations in nyctalopin have been reported to cause congenital stationary night blindness 1 (CSNB1) and isolated myopia without night blindness, making it a candidate gene for X-linked myopia disorder [107].

1.2.7.1 Chondroadherin

Chondroadherin also called as CHAD was first isolated from the bovine cartilage, and its chromosomal gene location is determined at a site 17q21.33 by isolation of a cosmid clone [108]. CHAD is known to mediate the adhesion of chondrocyte cartilage cells mainly, but it has been extracted in traces from small intestine, liver, skin, kidney, adrenal gland and pancreatic tissues [109]. The cDNA sequence of the human chondroadherin gene encodes a protein of 359 amino acids, and the first 21 amino acids represent a putative signal peptide sequence with eleven leucine-rich repeats flanked by cysteine-rich regions attached to the protein core forming a molecular mass of 36–38 kDa [108], Fig. 1.1. CHAD interacts with collagen, cell surface heparin sulphate proteoglycans and integrin $\alpha_2\beta_1$ receptors. The integrin-binding site is located at the C-terminal end and has been suggested to possess a therapeutic potential for treating osteoporosis [110].

CHAD is present in the cartilage tissue growth plate between the proliferative and hypertrophic zones and is distinctive from other SLRPs as it presents a double disulphide loop close to its C-terminus [111], while all other SLRP members have a single loop in this region. CHAD does not contain the N-terminal extension common to the other SLRPs and exists without any post-translational glycosylation sites [112]. CHAD binds with high affinity at two distinct sites of the triple helical collagen structure and mediates adhesion of isolated chondrocytes by two mechanisms: binding with the integrin $\alpha 2\beta 1$ receptor that mediates signalling between chondrocytes and their extracellular matrix; the other interaction is between the C-terminus chondroadherin sequence and cell surface proteoglycans such as syndecans that can also act as receptors. However, the precise region where chondroadherin interacts with the integrin is not completely elucidated [110]. CHAD interacts with the N- and C-terminus globular domains of collagen type VI; thus, they may provide mechanisms for regulating cellular activities in cartilage and bone tissue homeostasis [113]. Bone CHAD is known to promote attachment of osteoblastic cells and binds with high affinity to collagen types I and II [114].

CHAD can induce cell spreading with phorbol ester treatment to activate protein kinase C pathways necessary to initiate signals for cell spreading, forming focal adhesion kinases and stress fibres [115], and furthermore this process is suggested to involve integrin and syndecan receptors at the cell surface.

1.2.7.2 Tsukushi

Tsukushi also known as TSKU/TSK, E2-induced gene protein 4 (E2IG4) or leucine-rich repeat containing protein 54 (LRRC54) is a unique member of class IV SLRP family that shares functional properties with class I SLRPs [3]. The gene encoding TSK is located at chromosome 11q13.5 site with four exons. The distinct feature of TSK is the presence of twelve leucine-rich repeats with a cysteine-rich cluster at the N- and C-terminus [17, 116]. TSK was identified first by signal sequence trap screening using chick lens where it was demonstrated that the C-terminus of TSK migrates as a core protein band of 47 kDa in SDS-PAGE, and treatment with N-glycosidase-F (enzymes) reduces its size to 41 kDa indicating the existence of N-glycosylation [117]. In non-reducing condition (without enzymatic action), TSK migrates as a 90 kDa protein core, suggesting TSK forms homodimers in natural conditions. In chick embryogenesis model, C-terminus of TSK acts as an essential bone morphogenetic protein (BMP) inhibitor in the developing neural plate [118]. Further in vivo and in vitro investigations on mice show that TSK controls bone elongation and bone mass density by modulating growth plate chondrocyte function and consequently impacts the overall body size [27]. As longitudinal bone growth is determined by the rate of growth plate chondrogenesis, disorders involving TSK lead to severe skeletal dysplasia, short or tall stature, bowlegs and even club feet [119].

TSK is also an important component of the molecular pathways controlling retinal and peripheral eye development and acts as a Wnt signalling inhibitor which competes with Wnt2B for binding to Wnt receptor FZD4 and represses Wnt2B-dependent development of the eyes [120]. TSK is further shown to control the differentiation of the hair cycle by directly binding to TGF-β1 activity, and TSK depletion results in delay of the hair cycle with low expressions levels of TGF-β1 and pSmad2/3 [121].

In adult mice, TSK is localised in the organ of corti, spiral ganglion cells and the stria vascularis of the inner ear, and its loss causes changes the expression of Sox-2 and BMP-4 genes that drive the differentiation of the inner ear hair cells including stereocilia formations [122]. The role of TSK during wound healing in heterozygous, TSK-lacZ knock-in mice reveals a sequential pattern of TSK expression from macrophages and myofibroblasts. In vitro cell culture experiments suggest that TSK controls macrophage function and myofibroblast differentiation by inhibiting TGF-β1 secreted from macrophages facilitating wound healing by maintaining inflammatory cell quiescence [123]. TSK is produced and secreted by the liver, and expression of hepatic (liver) TSK is elevated in various obese mouse models. In nonalcoholic fatty liver disease (NAFLD), endoplasmic reticulum stress and inflammation promotes the release of TSK. Thus, in human hepatic tissue, TSK expression correlates with steatosis, and high plasma levels of TSK are reported in patients suffering from acetaminophen-induced acute liver failure (ALF). In mice model, TSK does not affect NAFLD development and progression but plays a role in the regulation of systemic cholesterol homeostasis. Thus, TSK lowers circulating HDL-cholesterol levels, reduces cholesterol efflux capacity and decreases cholesterol conversion to bile acids in the liver. Hepatic TSK is also identified as a blood biomarker of liver stress linked to NAFLD development with atherogenic dyslipidemia and atherosclerosis [124, 125].

1.2.7.3 Nyctalopin

Nyctalopin symbolised as NYX in humans encodes a gene protein located at chromosome Xp11.4 with three exons, a deduced protein of 481-amino-acid residue and a core protein size of 52 kDa [126, 127]. It is the only SLRP member with glycosylphosphatidylinositol (GPI)-anchored extracellular protein at the C-terminus and eleven typical leucine-rich repeats at the N-terminus. It is mainly expressed in the retina of the eyeball including tissues such as the brain, kidney, lungs, spleen and male testis in traces [128]. Within the retina of the eyes, NYX is mainly expressed in the inner segment of photoreceptors, outer and inner nuclear layers and the ganglion cell layer [129]. Defects in NYX gene protein expression cause congenital stationary night blindness-1 (CSNB1), also called as X-linked congenital stationary night blindness (XLCSNB) [130]. CSNB1 is a rare inherited retinal disorder characterised by impaired scotopic vision, myopia, hyperopia, nystagmus and reduced visual acuity. More than 50 mutations in NYX gene have been reported, and these include missense and nonsense mutations, splicing site mutations and insertions and deletions of genes coding for NYX [131]. However, the exact molecular pathogenesis mechanism of CSNB caused by NYX mutations remains to be fully elucidated. It has been suggested that nyctalopin may be involved in keeping the localisation of the transient receptor potential membrane-1(TRPM1) in the dendritic tips of the optic nerve bipolar cells in the eyes [132]. Nyctalopin may interact directly with TRPM1 or together with leucine-rich repeat Ig-like transmembrane domains-3 (LRIT3, another LRR family member) to keep the signalling channel proteins in the right place. Dislocation of TRPM1 lead to abnormal function of bipolar cells, resulting in altered signal transmission from the photoreceptors. Mutations in these three interacted proteins (nyctalopin, LRIT3 and TRPM1) have been documented to be associated with the complete form of congenital stationary night blindness [133].

1.2.8 Class V SLRPs

Class V SLRPs are non-canonical, with two members podocan and podocan-like protein-1 that exhibit properties such as collagen binding similar to the other SLRP members. Podocan comprises of twenty leucine-rich repeats attached to the N-terminus, and unlike the other SLRPs podocan lacks C-terminal cysteine residue which is replaced by a distinct acidic domain [134]. Podocan-like protein-1 is the other member of class V SLRP with a unique cysteine-rich N-terminus cluster, twenty one leucine-rich repeat and one putative N-glycosylation site [135].

1.2.8.1 Podocan

Podocan also known as PCAN, small leucine-rich repeat-5A (SLRR5A) or PODN was first isolated from human immunodeficiency virus transgenic mouse podocytes. The human podocan gene is located at chromosome 1p32.3 site, and it is a 613-amino-acid secreted protein containing 20 LRRs (leucine-rich repeats), a unique N-terminus cysteine-rich cluster and a highly acidic C-terminus domain [136]. Podocan protein has a core size of 66 kDa, Fig. 1.1, and is considered as a glycoprotein containing N-linked oligosaccharides that may be involved in cellular regulation of cardiovascular tissues [26] and are mainly expressed in the kidney, heart, liver, pancreas, urinary bladder, pineal gland and vascular smooth muscle cells. Podocan binds to type I collagen and can inhibit cell growth via the p-21 pathway and cell migration by RhoA activity. The podocan sequence can contain one consensus glycosaminoglycan side chain attached at serine–glycine motif in its N-terminus, and it typically lacks the C-terminus cysteine residues involved in disulphide formation [136].

1.2.8.2 Podocan-Like Protein-1

Podocan-like protein-1, also known as PODNL-1, was first identified in the bone matrix by bioinformatics and computational screening method [135]. It is an eighteen-amino-acid-long signal peptide with a unique cysteine-rich N-terminal cluster, a potential N-glycosylation site and twenty one leucine-rich repeats (LRRs). The PODNL-1 gene is located at chromosome 19p13.12 site and has nine exons; it is structurally similar to podocan, the other class V SLRP member. It is highly expressed in calvaria of the skull, mineralised tissues, lungs, kidney, urinary bladder and skin in traces [135, 137]. In bone osteoblastic cells, high expression level of PODNL-1was observed after matrix mineralisation in vitro. PODNL-1 is also considered to be a secreted glycoprotein with a core size of 62 kDa that undergoes post-translational modification with sugar moieties. PODNL-1 has been found to be highly expressed in glioma, and PODNL-1 knockdown mouse model suppresses glioma cell proliferation, migration and apoptosis partially via the Akt/mTOR pathway [138]. PODNL-1 is also suggested to be a novel biomarker and a potential therapeutic target for glioma; however, more details are yet to be fully elucidated.

References

1. Lazarides E. Modern cell biology: molecular cell biology. Science. 1986;234(4782):1448.
2. Luxenburg C, Geiger B. Multiscale view of cytoskeletal mechanoregulation of cell and tissue polarity. Handb Exp Pharmacol. 2017;235:263–84.
3. Dai G, Sun Y, Wei R, Xi L. Small leucine-rich proteoglycan. Curr Issues Mol Biol. 2023;45(7):6116–39.
4. Geiger B, Yamada KM. Molecular architecture and function of matrix adhesions. Cold Spring Harb Perspect Biol. 2011;3(5):a005033.
5. Iozzo RV. The family of the small leucine-rich proteoglycans: key regulators of matrix assembly and cellular growth. Crit Rev Biochem Mol Biol. 1997;32(2):141–74.
6. Moreth K, Iozzo RV, Schaefer L. Small leucine-rich proteoglycans orchestrate receptor crosstalk during inflammation. Cell Cycle. 2012;11(11):2084–91.
7. Schaefer L, Tsalastra W, Babelova A, Baliova M, Minnerup J, Sorokin L, et al. Decorin-mediated regulation of fibrillin-1 in the kidney involves the insulin-like growth factor-I receptor and mammalian target of rapamycin. Am J Pathol. 2007;170(1):301–15.
8. Iozzo RV. Matrix proteoglycans: from molecular design to cellular function. Annu Rev Biochem. 1998;67:609–52.
9. Brandan E, Retamal C, Cabello-Verrugio C, Marzolo MP. The low density lipoprotein receptor-related protein functions as an endocytic receptor for decorin. J Biol Chem. 2006;281(42):31562–71.
10. Iozzo RV, Schaefer L. Proteoglycans in health and disease: novel regulatory signaling mechanisms evoked by the small leucine-rich proteoglycans. FEBS J. 2010;277(19):3864–75.
11. Zeltz C, Brézillon S, Käpylä J, Eble JA, Bobichon H, Terryn C, et al. Lumican inhibits cell migration through $\alpha 2\beta 1$ integrin. Exp Cell Res. 2010;316(17):2922–31.
12. Goldoni S, Humphries A, Nyström A, Sattar S, Owens RT, McQuillan DJ, et al. Decorin is a novel antagonistic ligand of the met receptor. J Cell Biol. 2009;185(4):743–54.
13. Merline R, Schaefer RM, Schaefer L. The matricellular functions of small leucine-rich proteoglycans (SLRPs). J Cell Commun Signaling. 2009;3(3–4):323–35.
14. Chen S, Birk DE. The regulatory roles of small leucine-rich proteoglycans in extracellular matrix assembly. FEBS J. 2013;280(10):2120–37.
15. Horiguchi M, Ota M, Rifkin DB. Matrix control of transforming growth factor-β function. J Biochem. 2012;152(4):321–9.
16. Young MF, Bi Y, Ameye L, Chen XD. Biglycan knockout mice: new models for musculoskeletal diseases. Glycoconj J. 2002;19(4–5):257–62.
17. Schaefer L, Iozzo RV. Biological functions of the small leucine-rich proteoglycans: from genetics to signal transduction. J Biol Chem. 2008;283(31):21305–9.
18. McEwan PA, Scott PG, Bishop PN, Bella J. Structural correlations in the family of small leucine-rich repeat proteins and proteoglycans. J Struct Biol. 2006;155(2):294–305.
19. Solis-Hernandez MP, Martín C, García B, Pérez-López N, García-Mesa Y, González-Fernández S, et al. The genes encoding small leucine-rich proteoglycans undergo differential expression alterations in colorectal cancer, depending on tumor location. Cells. 2021;10(8):2002.
20. Pietraszek-Gremplewicz K, Karamanou K, Niang A, Dauchez M, Belloy N, Maquart FX, et al. Small leucine-rich proteoglycans and matrix metalloproteinase-14: key partners? Matrix Biol. 2019;75–76:271–85.
21. Matsushima N, Miyashita H, Kretsinger RH. Sequence features, structure, ligand interaction, and diseases in small leucine rich repeat proteoglycans. J Cell Commun Signaling. 2021;15(4):519–31.
22. Matsushima N, Tanaka T, Enkhbayar P, Mikami T, Taga M, Yamada K, et al. Comparative sequence analysis of leucine-rich repeats (LRRs) within vertebrate toll-like receptors. BMC Genomics. 2007;8:124.

23. Zappia J, Joiret M, Sanchez C, Lambert C, Geris L, Muller M, et al. From translation to protein degradation as mechanisms for regulating biological functions: a review on the SLRP family in skeletal tissues. Biomol Ther. 2020;10(1):80.
24. Iozzo RV. The biology of the small leucine-rich proteoglycans. Functional network of interactive proteins. J Biol Chem. 1999;274(27):18843–6.
25. Matsushima N, Miyashita H, Mikami T, Kuroki Y. A nested leucine rich repeat (LRR) domain: the precursor of LRRs is a ten or eleven residue motif. BMC Microbiol. 2010;10:235.
26. Iozzo RV, Schaefer L. Proteoglycan form and function: a comprehensive nomenclature of proteoglycans. Matrix Biol. 2015;42:11–55.
27. Yano K, Washio K, Tsumanuma Y, Yamato M, Ohta K, Okano T, et al. The role of Tsukushi (TSK), a small leucine-rich repeat proteoglycan, in bone growth. Regen Ther. 2017;7:98–107.
28. Chen XD, Fisher LW, Robey PG, Young MF. The small leucine-rich proteoglycan biglycan modulates BMP-4-induced osteoblast differentiation. FASEB J. 2004;18(9):948–58.
29. Dolan J, Walshe K, Alsbury S, Hokamp K, O'Keeffe S, Okafuji T, et al. The extracellular leucine-rich repeat superfamily; a comparative survey and analysis of evolutionary relationships and expression patterns. BMC Genomics. 2007;8:320.
30. Scott PG, McEwan PA, Dodd CM, Bergmann EM, Bishop PN, Bella J. Crystal structure of the dimeric protein core of decorin, the archetypal small leucine-rich repeat proteoglycan. Proc Natl Acad Sci USA. 2004;101(44):15633–8.
31. Neill T, Schaefer L, Iozzo RV. Decorin: a guardian from the matrix. Am J Pathol. 2012;181(2):380–7.
32. Orgel JP, Eid A, Antipova O, Bella J, Scott JE. Decorin core protein (decoron) shape complements collagen fibril surface structure and mediates its binding. PLoS One. 2009;4(9):e7028.
33. McBride OW, Fisher LW, Young MF. Localization of PGI (biglycan, BGN) and PGII (decorin, DCN, PG-40) genes on human chromosomes Xq13-qter and 12q, respectively. Genomics. 1990;6(2):219–25.
34. Choi HU, Johnson TL, Pal S, Tang LH, Rosenberg L, Neame PJ. Characterization of the dermatan sulfate proteoglycans, DS-PGI and DS-PGII, from bovine articular cartilage and skin isolated by octyl-sepharose chromatography. J Biol Chem. 1989;264(5):2876–84.
35. Roughley PJ, White RJ, Magny MC, Liu J, Pearce RH, Mort JS. Non-proteoglycan forms of biglycan increase with age in human articular cartilage. Biochem J. 1993;295(Pt 2):421–6.
36. Schaefer L, Schaefer RM. Proteoglycans: from structural compounds to signaling molecules. Cell Tissue Res. 2010;339(1):237–46.
37. Nastase MV, Young MF, Schaefer L. Biglycan: a multivalent proteoglycan providing structure and signals. J Histochem Cytochem. 2012;60(12):963–75.
38. Shainer R, Kram V, Kilts TM, Li L, Doyle AD, Shainer I, et al. Biglycan regulates bone development and regeneration. Front Physiol. 2023;14:1119368.
39. Henry SP, Takanosu M, Boyd TC, Mayne PM, Eberspaecher H, Zhou W, et al. Expression pattern and gene characterization of asporin. A newly discovered member of the leucine-rich repeat protein family. J Biol Chem. 2001;276(15):12212–21.
40. Fan R, Yan X, Zhang W. Relationship between asporin and extracellular matrix behavior: a literature review. Medicine (Baltimore). 2022;101(51):e32490.
41. Ikegawa S. Expression, regulation and function of asporin, a susceptibility gene in common bone and joint diseases. Curr Med Chem. 2008;15(7):724–8.
42. Nakamura T, Shi D, Tzetis M, Rodriguez-Lopez J, Miyamoto Y, Tsezou A, et al. Meta-analysis of association between the ASPN D-repeat and osteoarthritis. Hum Mol Genet. 2007;16(14):1676–81.
43. Önnerfjord P, Khabut A, Reinholt FP, Svensson O, Heinegård D. Quantitative proteomic analysis of eight cartilaginous tissues reveals characteristic differences as well as similarities between subgroups. J Biol Chem. 2012;287(23):18913–24.
44. Nakajima M, Kizawa H, Saitoh M, Kou I, Miyazono K, Ikegawa S. Mechanisms for asporin function and regulation in articular cartilage. J Biol Chem. 2007;282(44):32185–92.

45. Maris P, Blomme A, Palacios AP, Costanza B, Bellahcène A, Bianchi E, et al. Asporin is a fibroblast-derived TGF-β1 inhibitor and a tumor suppressor associated with good prognosis in breast cancer. PLoS Med. 2015;12(9):e1001871.
46. Nishiu J, Tanaka T, Nakamura Y. Identification of a novel gene (ECM2) encoding a putative extracellular matrix protein expressed predominantly in adipose and female-specific tissues and its chromosomal localization to 9q22.3. Genomics. 1998;52(3):378–81.
47. Wang X, Shen Y, Shang M, Liu X, Munn LL. Endothelial mechanobiology in atherosclerosis. Cardiovasc Res. 2023;119(8):1656–75.
48. Karamanos NK, Theocharis AD, Piperigkou Z, Manou D, Passi A, Skandalis SS, et al. A guide to the composition and functions of the extracellular matrix. FEBS J. 2021;288(24):6850–912.
49. Zheng Z, Granado HS, Li C. Fibromodulin, a Multifunctional Matricellular Modulator. J Dent Res. 2023;102(2):125–34.
50. Paluru PC, Scavello GS, Ganter WR, Young TL. Exclusion of lumican and fibromodulin as candidate genes in MYP3 linked high grade myopia. Mol Vis. 2004;10:917–22.
51. Zheng Z, Li C, Ha P, Chang GX, Yang P, Zhang X, et al. CDKN2B upregulation prevents teratoma formation in multipotent fibromodulin-reprogrammed cells. J Clin Invest. 2019;129(8):3236–51.
52. Lauder RM, Huckerby TN, Nieduszynski IA. The structure of the keratan sulphate chains attached to fibromodulin isolated from articular cartilage. Eur J Biochem. 1996;242(2):402–9.
53. Oldberg A, Antonsson P, Lindblom K, Heinegård D. A collagen-binding 59-kd protein (fibromodulin) is structurally related to the small interstitial proteoglycans PG-S1 and PG-S2 (decorin). EMBO J. 1989;8(9):2601–4.
54. Zheng Z, Zhang X, Dang C, Beanes S, Chang GX, Chen Y, et al. Fibromodulin is essential for fetal-type Scarless cutaneous wound healing. Am J Pathol. 2016;186(11):2824–32.
55. Zheng Z, James AW, Li C, Jiang W, Wang JZ, Chang GX, et al. Fibromodulin reduces scar formation in adult cutaneous wounds by eliciting a fetal-like phenotype. Signal Transduct Target Ther. 2017;2:17050.
56. Grover J, Chen XN, Korenberg JR, Roughley PJ. The human lumican gene. Organization, chromosomal location, and expression in articular cartilage. J Biol Chem. 1995;270(37):21942–9.
57. Wu F, Vij N, Roberts L, Lopez-Briones S, Joyce S, Chakravarti S. A novel role of the lumican core protein in bacterial lipopolysaccharide-induced innate immune response. J Biol Chem. 2007;282(36):26409–17.
58. Brézillon S, Pietraszek K, Maquart FX, Wegrowski Y. Lumican effects in the control of tumour progression and their links with metalloproteinases and integrins. FEBS J. 2013;280(10):2369–81.
59. Melching LI, Roughley PJ. Modulation of keratan sulfate synthesis on lumican by the action of cytokines on human articular chondrocytes. Matrix Biol. 1999;18(4):381–90.
60. Kao WW, Liu CY. Roles of lumican and keratocan on corneal transparency. Glycoconj J. 2002;19(4–5):275–85.
61. Onda M, Ishiwata T, Kawahara K, Wang R, Naito Z, Sugisaki Y. Expression of lumican in thickened intima and smooth muscle cells in human coronary atherosclerosis. Exp Mol Pathol. 2002;72(2):142–9.
62. Krishnan A, Li X, Kao WY, Viker K, Butters K, Masuoka H, et al. Lumican, an extracellular matrix proteoglycan, is a novel requisite for hepatic fibrosis. Lab Investig. 2012;92(12):1712–25.
63. Karamanou K, Franchi M, Piperigkou Z, Perreau C, Maquart FX, Vynios DH, et al. Lumican effectively regulates the estrogen receptors-associated functional properties of breast cancer cells, expression of matrix effectors and epithelial-to-mesenchymal transition. Sci Rep. 2017;7:45138.
64. Nikitovic D, Katonis P, Tsatsakis A, Karamanos NK, Tzanakakis GN. Lumican, a small leucine-rich proteoglycan. IUBMB Life. 2008;60(12):818–23.
65. Chakravarti S, Magnuson T, Lass JH, Jepsen KJ, LaMantia C, Carroll H. Lumican regulates collagen fibril assembly: skin fragility and corneal opacity in the absence of lumican. J Cell Biol. 1998;141(5):1277–86.

66. Berdiaki A, Giatagana EM, Tzanakakis G, Nikitovic D. The landscape of small leucine-rich proteoglycan impact on cancer pathogenesis with a focus on Biglycan and Lumican. Cancers (Basel). 2023;15(14):3549.
67. Bengtsson E, Neame PJ, Heinegård D, Sommarin Y. The primary structure of a basic leucine-rich repeat protein, PRELP, found in connective tissues. J Biol Chem. 1995;270(43):25639–44.
68. Grover J, Chen XN, Korenberg JR, Recklies AD, Roughley PJ. The gene organization, chromosome location, and expression of a 55-kDa matrix protein (PRELP) of human articular cartilage. Genomics. 1996;38(2):109–17.
69. Bengtsson E, Mörgelin M, Sasaki T, Timpl R, Heinegård D, Aspberg A. The leucine-rich repeat protein PRELP binds perlecan and collagens and may function as a basement membrane anchor. J Biol Chem. 2002;277(17):15061–8.
70. Bengtsson E, Aspberg A, Heinegard D, Sommarin Y, Spillmann D. The amino-terminal part of PRELP binds to heparin and heparan sulfate. J Biol Chem. 2000;275(52):40695–702.
71. Johansen JS, Williamson MK, Rice JS, Price PA. Identification of proteins secreted by human osteoblastic cells in culture. J Bone Miner Res. 1992;7(5):501–12.
72. Rucci N, Rufo A, Alamanou M, Capulli M, Del Fattore A, Ahrman E, et al. The glycosaminoglycan-binding domain of PRELP acts as a cell type-specific NF-kappaB inhibitor that impairs osteoclastogenesis. J Cell Biol. 2009;187(5):669–83.
73. Aspberg A. The different roles of aggrecan interaction domains. J Histochem Cytochem. 2012;60(12):987–96.
74. Grover J, Roughley PJ. Characterization of the human proline/arginine-rich end leucine-rich repeat protein (PRELP) gene promoter and identification of a repressor element. Biochem J. 1998;336(Pt 1):77–82.
75. Tasheva ES, Funderburgh JL, Funderburgh ML, Corpuz LM, Conrad GW. Structure and sequence of the gene encoding human keratocan. DNA Seq. 1999;10(1):67–74.
76. Corpuz LM, Funderburgh JL, Funderburgh ML, Bottomley GS, Prakash S, Conrad GW. Molecular cloning and tissue distribution of keratocan. Bovine corneal keratan sulfate proteoglycan 37A. J Biol Chem. 1996;271(16):9759–63.
77. Igwe JC, Gao Q, Kizivat T, Kao WW, Kalajzic I. Keratocan is expressed by osteoblasts and can modulate osteogenic differentiation. Connect Tissue Res. 2011;52(5):401–7.
78. Caterson B, Melrose J. Keratan sulfate, a complex glycosaminoglycan with unique functional capability. Glycobiology. 2018;28(4):182–206.
79. Funderburgh JL. Keratan sulfate: structure, biosynthesis, and function. Glycobiology. 2000;10(10):951–8.
80. Lin W, Gao L, Jiang W, Niu C, Yuan K, Hu X, et al. The role of osteomodulin on osteo/odontogenic differentiation in human dental pulp stem cells. BMC Oral Health. 2019;19(1):22.
81. Wendel M, Sommarin Y, Heinegård D. Bone matrix proteins: isolation and characterization of a novel cell-binding keratan sulfate proteoglycan (osteoadherin) from bovine bone. J Cell Biol. 1998;141(3):839–47.
82. Lin W, Zhu X, Gao L, Mao M, Gao D, Huang Z. Osteomodulin positively regulates osteogenesis through interaction with BMP2. Cell Death Dis. 2021;12(2):147.
83. Petersson U, Hultenby K, Wendel M. Identification, distribution and expression of osteoadherin during tooth formation. Eur J Oral Sci. 2003;111(2):128–36.
84. Nikdin H, Olsson ML, Hultenby K, Sugars RV. Osteoadherin accumulates in the predentin towards the mineralization front in the developing tooth. PLoS One. 2012;7(2):e31525.
85. Friedman JS, Faucher M, Hiscott P, Biron VL, Malenfant M, Turcotte P, et al. Protein localization in the human eye and genetic screen of opticin. Hum Mol Genet. 2002;11(11):1333–42.
86. Reardon AJ, Le Goff M, Briggs MD, McLeod D, Sheehan JK, Thornton DJ, et al. Identification in vitreous and molecular cloning of opticin, a novel member of the family of leucine-rich repeat proteins of the extracellular matrix. J Biol Chem. 2000;275(3):2123–9.
87. Takanosu M, Boyd TC, Le Goff M, Henry SP, Zhang Y, Bishop PN, et al. Structure, chromosomal location, and tissue-specific expression of the mouse opticin gene. Invest Ophthalmol Vis Sci. 2001;42(10):2202–10.

88. Monfort J, Tardif G, Roughley P, Reboul P, Boileau C, Bishop PN, et al. Identification of opticin, a member of the small leucine-rich repeat proteoglycan family, in human articular tissues: a novel target for MMP-13 in osteoarthritis. Osteoarthr Cartil. 2008;16(7):749–55.
89. Le Goff MM, Sutton MJ, Slevin M, Latif A, Humphries MJ, Bishop PN. Opticin exerts its anti-angiogenic activity by regulating extracellular matrix adhesiveness. J Biol Chem. 2012;287(33):28027–36.
90. Mikaelsson E, Osterborg A, Tahmasebi Fard Z, Mahmoudi A, Mahmoudian J, Jeddi-Tehrani M, et al. Opticin, a small leucine-rich proteoglycan, is uniquely expressed and translocated to the nucleus of chronic lymphocytic leukemia cells. Exp Hematol Oncol. 2013;2(1):23.
91. Le Goff MM, Bishop PN. Adult vitreous structure and postnatal changes. Eye (Lond). 2008;22(10):1214–22.
92. Le Goff MM, Hindson VJ, Jowitt TA, Scott PG, Bishop PN. Characterization of opticin and evidence of stable dimerization in solution. J Biol Chem. 2003;278(46):45280–7.
93. Tasheva ES, Pettenati M, Von Kap-Her C, Conrad GW. Assignment of mimecan gene (OGN) to human chromosome band 9q22 by in situ hybridization. Cytogenet Cell Genet. 2000;88(3–4):326–7.
94. Tanaka K, Matsumoto E, Higashimaki Y, Katagiri T, Sugimoto T, Seino S, et al. Role of osteoglycin in the linkage between muscle and bone. J Biol Chem. 2012;287(15):11616–28.
95. Tasheva ES, Funderburgh ML, McReynolds J, Funderburgh JL, Conrad GW. The bovine mimecan gene. Molecular cloning and characterization of two major RNA transcripts generated by alternative use of two splice acceptor sites in the third exon. J Biol Chem. 1999;274(26):18693–701.
96. Funderburgh JL, Corpuz LM, Roth MR, Funderburgh ML, Tasheva ES, Conrad GW. Mimecan, the 25-kDa corneal keratan sulfate proteoglycan, is a product of the gene producing osteoglycin. J Biol Chem. 1997;272(44):28089–95.
97. Nulali J, Zhan M, Zhang K, Tu P, Liu Y, Song H. Osteoglycin: an ECM factor regulating fibrosis and tumorigenesis. Biomol Ther. 2022;12(11):1674.
98. Deckx S, Heymans S, Papageorgiou AP. The diverse functions of osteoglycin: a deceitful dwarf, or a master regulator of disease? FASEB J. 2016;30(8):2651–61.
99. Shanahan CM, Cary NR, Osbourn JK, Weissberg PL. Identification of osteoglycin as a component of the vascular matrix. Differential expression by vascular smooth muscle cells during neointima formation and in atherosclerotic plaques. Arterioscler Thromb Vasc Biol. 1997;17(11):2437–47.
100. Dimberg A. Osteoglycin - a switch from angiogenesis to T-cell recruitment? EBioMedicine. 2018;35:22–3.
101. Deere M, Dieguez JL, Yoon SJ, Hewett-Emmett D, de la Chapelle A, Hecht JT. Genomic characterization of human DSPG3. Genome Res. 1999;9(5):449–56.
102. Johnson HJ, Rosenberg L, Choi HU, Garza S, Höök M, Neame PJ. Characterization of epiphycan, a small proteoglycan with a leucine-rich repeat core protein. J Biol Chem. 1997;272(30):18709–17.
103. Deere M, Johnson J, Garza S, Harrison WR, Yoon SJ, Elder FF, et al. Characterization of human DSPG3, a small dermatan sulfate proteoglycan. Genomics. 1996;38(3):399–404.
104. Nuka S, Zhou W, Henry SP, Gendron CM, Schultz JB, Shinomura T, et al. Phenotypic characterization of epiphycan-deficient and epiphycan/biglycan double-deficient mice. Osteoarthr Cartil. 2010;18(1):88–96.
105. Kim MJ, Frausto RF, Rosenwasser GO, Bui T, Le DJ, Stone EM, et al. Posterior amorphous corneal dystrophy is associated with a deletion of small leucine-rich proteoglycans on chromosome 12. PLoS One. 2014;9(4):e95037.
106. O'Connor E, Eisenhaber B, Dalley J, Wang T, Missen C, Bulleid N, et al. Species specific membrane anchoring of nyctalopin, a small leucine-rich repeat protein. Hum Mol Genet. 2005;14(13):1877–87.
107. Zhang Q, Xiao X, Li S, Jia X, Yang Z, Huang S, et al. Mutations in NYX of individuals with high myopia, but without night blindness. Mol Vis. 2007;13:330–6.

108. Grover J, Chen XN, Korenberg JR, Roughley PJ. The structure and chromosome location of the human chondroadherin gene (Chad). Genomics. 1997;45(2):379–85.
109. Landgren C, Beier DR, Fässler R, Heinegård D, Sommarin Y. The mouse chondroadherin gene: characterization and chromosomal localization. Genomics. 1998;47(1):84–91.
110. Haglund L, Tillgren V, Addis L, Wenglén C, Recklies A, Heinegård D. Identification and characterization of the integrin alpha2beta1 binding motif in chondroadherin mediating cell attachment. J Biol Chem. 2011;286(5):3925–34.
111. Hessle L, Stordalen GA, Wenglén C, Petzold C, Tanner E, Brorson SH, et al. The skeletal phenotype of chondroadherin deficient mice. PLoS One. 2014;8(6):e63080.
112. Paracuellos P, Kalamajski S, Bonna A, Bihan D, Farndale RW, Hohenester E. Structural and functional analysis of two small leucine-rich repeat proteoglycans, fibromodulin and chondroadherin. Matrix Biol. 2017;63:106–16.
113. Camper L, Heinegârd D, Lundgren-Akerlund E. Integrin alpha2beta1 is a receptor for the cartilage matrix protein chondroadherin. J Cell Biol. 1997;138(5):1159–67.
114. Mizuno M, Fujisawa R, Kuboki Y. Bone chondroadherin promotes attachment of osteoblastic cells to solid-state substrates and shows affinity to collagen. Calcif Tissue Int. 1996;59(3):163–7.
115. Haglund L, Tillgren V, Önnerfjord P, Heinegård D. The C-terminal peptide of chondroadherin modulates cellular activity by selectively binding to heparan sulfate chains. J Biol Chem. 2013;288(2):995–1008.
116. Li Y, Jin L, Yan J, Huang Y, Zhang H, Zhang R, et al. Tsukushi and TSKU genotype in obesity and related metabolic disorders. J Endocrinol Investig. 2021;44(12):2645–54.
117. Ohta K, Lupo G, Kuriyama S, Keynes R, Holt CE, Harris WA, et al. Tsukushi functions as an organizer inducer by inhibition of BMP activity in cooperation with chordin. Dev Cell. 2004;7(3):347–58.
118. Ahmad SAI, Anam MB, Ito N, Ohta K. Involvement of Tsukushi in diverse developmental processes. J Cell Commun Signaling. 2018;12(1):205–10.
119. Nikitovic D, Aggelidakis J, Young MF, Iozzo RV, Karamanos NK, Tzanakakis GN. The biology of small leucine-rich proteoglycans in bone pathophysiology. J Biol Chem. 2012;287(41):33926–33.
120. Ohta K, Ito A, Kuriyama S, Lupo G, Kosaka M, Ohnuma S, et al. Tsukushi functions as a Wnt signaling inhibitor by competing with Wnt2b for binding to transmembrane protein Frizzled4. Proc Natl Acad Sci USA. 2011;108(36):14962–7.
121. Niimori D, Kawano R, Felemban A, Niimori-Kita K, Tanaka H, Ihn H, et al. Tsukushi controls the hair cycle by regulating TGF-β1 signaling. Dev Biol. 2012;372(1):81–7.
122. Miwa T, Ohta K, Ito N, Hattori S, Miyakawa T, Takeo T, et al. Tsukushi is essential for the development of the inner ear. Mol Brain. 2020;13(1):29.
123. Niimori D, Kawano R, Niimori-Kita K, Ihn H, Ohta K. Tsukushi is involved in the wound healing by regulating the expression of cytokines and growth factors. J Cell Commun Signaling. 2014;8(3):173–7.
124. Mouchiroud M, Camiré É, Aldow M, Caron A, Jubinville É, Turcotte L, et al. The hepatokine Tsukushi is released in response to NAFLD and impacts cholesterol homeostasis. JCI Insight. 2019;4(15):e129492.
125. Lam S, Lee CH, Fong CHY, Wong Y, Shiu SWM, Mak LY, et al. Serum Tsukushi level is associated with the severity of liver fibrosis independent of type 2 diabetes. J Clin Endocrinol Metab. 2024;109(3):e1048–e54.
126. Scanga HL, Liasis A, Pihlblad MS, Nischal KK. Related congenital stationary night blindness in two siblings due to probable maternal germline mosaicism. Ophthalmic Genet. 2021;42(5):588–92.
127. Gregg RG, Mukhopadhyay S, Candille SI, Ball SL, Pardue MT, McCall MA, et al. Identification of the gene and the mutation responsible for the mouse nob phenotype. Invest Ophthalmol Vis Sci. 2003;44(1):378–84.

128. Pesch K, Zeitz C, Fries JE, Münscher S, Pusch CM, Kohler K, et al. Isolation of the mouse nyctalopin gene nyx and expression studies in mouse and rat retina. Invest Ophthalmol Vis Sci. 2003;44(5):2260–6.
129. Hoon M, Okawa H, Della Santina L, Wong RO. Functional architecture of the retina: development and disease. Prog Retin Eye Res. 2014;42:44–84.
130. Bech-Hansen NT, Naylor MJ, Maybaum TA, Sparkes RL, Koop B, Birch DG, et al. Mutations in NYX, encoding the leucine-rich proteoglycan nyctalopin, cause X-linked complete congenital stationary night blindness. Nat Genet. 2000;26(3):319–23.
131. Pusch CM, Zeitz C, Brandau O, Pesch K, Achatz H, Feil S, et al. The complete form of X-linked congenital stationary night blindness is caused by mutations in a gene encoding a leucine-rich repeat protein. Nat Genet. 2000;26(3):324–7.
132. Cao Y, Posokhova E, Martemyanov KA. TRPM1 forms complexes with nyctalopin in vivo and accumulates in postsynaptic compartment of ON-bipolar neurons in mGluR6-dependent manner. J Neurosci. 2011;31(32):11521–6.
133. van Genderen MM, Bijveld MM, Claassen YB, Florijn RJ, Pearring JN, Meire FM, et al. Mutations in TRPM1 are a common cause of complete congenital stationary night blindness. Am J Hum Genet. 2009;85(5):730–6.
134. Ross MD, Bruggeman LA, Hanss B, Sunamoto M, Marras D, Klotman ME, et al. Podocan, a novel small leucine-rich repeat protein expressed in the sclerotic glomerular lesion of experimental HIV-associated nephropathy. J Biol Chem. 2003;278(35):33248–55.
135. Mochida Y, Kaku M, Yoshida K, Katafuchi M, Atsawasuwan P, Yamauchi M. Podocan-like protein: a novel small leucine-rich repeat matrix protein in bone. Biochem Biophys Res Commun. 2011;410(2):333–8.
136. Shimizu-Hirota R, Sasamura H, Kuroda M, Kobayashi E, Saruta T. Functional characterization of podocan, a member of a new class in the small leucine-rich repeat protein family. FEBS Lett. 2004;563(1–3):69–74.
137. Liu X, He Y, Tong Y, Huang Q, Shang H, Chen Z, et al. PODNL1 promotes cell migration and regulates the epithelial/mesenchymal transition process in bladder cancer. Biochem Biophys Res Commun. 2022;620:165–72.
138. Geng Y, Zuo P, Li XO, Zhang L. PODNL1 promotes cell proliferation and migration in glioma via regulating Akt/mTOR pathway. J Cancer. 2020;11(21):6234–42.

Diseases of the Matrix and the Role of SLRPs in Inflammatory Catabolism

2

Polly Lama and Michael A. Adams

Abstract

Small leucine-rich proteoglycan (SLRP) molecules are involved in the regulation of sterile inflammatory disorder as recent and past research spanning over two decades provide ample evidence on the role of various SLRPs and its fragments in tissue catabolism. As described in Chap. 1, SLRPs are ubiquitous, non-collagenous, secreted proteins with a core size of approximately 36–60 kDa presenting a characteristic leucine-rich repeat flanked with or without glycosylated side chains. The SLRP protein core undergoes post-translational modification during tissue stress, injury and pathological diseases breaking down the SLRP molecules in smaller fragments which are released into the extracellular matrix; these proteolytic fragments are able to interact with cellular receptors, cytokines, chemokines, growth factors, enzymes and many other matrix components. SLRPs can thus exist in a matrix-bound, sequestered or even a soluble form which reflects as the stage of the diseased matrix. Therefore, the main aim of this chapter is to elucidate in detail about the SLRP types that are most active in cancer and cartilage, renal, skin and other common tissue diseases compromising the extracellular matrix integrity. This chapter also explains about the dual role of SLRPs as an inhibitor and an activator of matrix metalloproteases.

P. Lama (✉)
Department of Anatomy, Sikkim Manipal Institute of Medical Sciences, Sikkim Manipal University, Gangtok, Sikkim, India
e-mail: polly.lama@smims.smu.edu.in

M. A. Adams
Centre for Applied Anatomy, University of Bristol, Bristol, UK

P. Lama (ed.), *The Biological Role of Small Leucine-Rich Proteoglycans (SLRPs) in Matrix Homeostasis and Diseases*,
https://doi.org/10.1007/978-981-96-7890-7_2

Keywords

Small leucine rich proteoglycans · Cancer · Inflammation · Bone · Cartilage · Tissues

2.1 SLRPs, Inflammation and Cancer

Cancer or malignant neoplasm involves uncontrolled cell division and growth that leads to the formation of tumours disrupting the normal structure and functions of the affected organ and the tissues [1]. In metastasis, cancer cells penetrate and infiltrate the surrounding normal tissues leading to the rise of new tumours [2]. Cancer cells can pass through the blood vessels to become circulating tumour cells, whereby they can metastasise to more distant organs or tissues of the body.

Cancer is a rising non-communicable disease among adults aged between 40 and 80 years and accounts for ten million deaths annually with lung cancer as the most common type of case followed by liver, stomach, colorectal and breast cancer [3]. The risk factors for cancer are attributed to environmental exposure such as smoking, nutrition, infection, stress and radiation, while a small 5–10% of the cases can occur due to genetic defects that develop and progress to induce multiple alterations such as activation of oncogenes and inactivation of tumour-suppressor genes (TSG) that can also arise through the aforementioned environmental factors [4].

Cancer cells create their own microenvironment that essentially affects themselves and the surrounding cells of the extracellular matrix [5]. The tumour microenvironment is composed of endothelial cells, pericytes, invading inflammatory cells, leucocytes, fibroblasts, host tissue-specific cells and multiple other matrix components forming the tumour stroma [6].

The extracellular matrix of the tumour-associated tissue is fundamentally different from normal tissue structure, as distinct morphological and molecular changes can promote cell transformation, tumour growth, motility and invasion, enabling metastatic dissemination to facilitate emergence of secondary tumours at distant sites [7]. One of the important mediators in tumour progression are the enzymes matrix metalloproteinases (MMPs), a family of calcium-dependent zinc-containing endopeptidases that promotes tumorigenesis by breaking down the extracellular matrix components allowing the cancer cells to invade the adjacent normal tissue [8].

During carcinogenesis, many growth factors, cytokines and enzymes are secreted from the malignant cells that further stimulate cancer cell growth and allow the surrounding stromal cells to secrete molecules that increase the permissiveness of tissues in tumour growth [9]. Thus, the extracellular matrix integrity is crucial in maintaining tissue homeostasis preventing invasion and metastasis of cancer cells, as both tumour and stromal cells are involved with the reorganisation of the matrix. In this process, SLRPs tend to act as an important regulator as their expression profile is markedly altered in the tumour microenvironment [10]. Modified secretion of SLRPs, and matrix-accumulated proteoglycans, in stromal cell membranes can affect survival, signalling, growth, adhesion and migration of the cancer cells.

Almost eight members of SLRPs family have been found to be implicated in cancer, and a characteristic feature of these SLRPs is their ability to interact with extracellular receptors such as the tyrosine kinase and toll-like receptors facilitated by the bare β-sheets present on the concave surface of their leucine-rich repeats (LRRs). This interaction initiates downstream signalling that regulates cell–matrix function especially proliferation and inflammation.

Decorin (DCN), biglycan (BGN) and asporin (ASPN) owing to their small size and molecular fragments are shown to be considerably involved in cancer pathogenesis [11] and are described in detail in the following subsections.

2.1.1 Decorin

Decorin is associated with neo-angiogenesis, a process involving the formation of new blood vessels in cancer tissue, and decorin is not expressed by the quiescent (inactive) endothelial vascular cells [12]. Decorin is synthesised during angiogenic morphogenesis along with the inflammatory mediator interleukins 6 and 10 [13] playing a stimulatory and inhibitory role, while being differentially expressed in human benign and malignant vascular tumours [14]. In Kaposi's sarcoma and angiosarcoma, representing malignant vascular neoplasms, decorin expression is completely lacking, whereas in benign vascular tumours, namely, haemangiomas, where capillary growth has ceased, decorin is expressed in detectable amounts. Decorin is involved in the regulation of tumour cells through transforming growth factor beta (TGF-β) signalling pathway inhibiting growth [15] by downregulating the epidermal growth factor (EGF) receptor [16] and the receptor tyrosine kinase of the EGFR family [17]. Past studies also show molecular fragments of decorin exhibit antiangiogenic activities, possibly through the ability of these fragments supressing vascular endothelial growth factor (VEGF)-induced phosphorylation and assembly of focal adhesion kinase complexes [18]. Overexpression of decorin inhibits corneal neo-vascularisation by downregulating the proangiogenic VEGF [19], and gene defects in decorin cause intestinal tumour formation by altering intestinal cell maturation process [20]. Mice with a double deficiency of decorin inhibits p53 signalling pathway and prematurely develop aggressive lymphomas [21].

Hepatitis B virus (HBV) is a potent risk factor inducing the occurrence and development of liver cancer [22], and *in hepatocellular carcinoma* (HCC) decorin mRNA expression tends to be significantly lower in patient tumour tissues than the healthy control, where the HBV encoded oncogene-X (HBx) inhibits the transcription of decorin by blocking p53 activity [23]. Thus, overexpression of decorin inhibits the multiplication of HCC cells, while knockdown enhances uncontrolled growth, and decorin acts by competitively binding to c-Met to inhibit hepatocellular growth factor (HGF) signalling pathway leading to the development of carcinoma [24]. Thus, DCN-derived antitumour peptides (DCN-Ps) have been shown to competitively bind with c-Met and are proposed to act as a novel compound for developing therapeutic drugs against HCC [23].

In breast carcinoma cells decorin activates mitophagy by binding to the Met receptor that activates peroxisome proliferator receptor-γ-coactivator-1α (PGC-1α), leading to the accumulation of mitostatin, a decorin-induced tumour suppressor gene [25].

In pancreatic cancer decorin is widely localised in the tissue and shows higher expression than para-cancerous tissue during stage I of the disease. In stage II, the distribution and expression of decorin is similar in the cancerous and the para-cancerous pancreatic tissue, while in stage III the expression is more pronounced in the para-cancerous tissue [26].

Decorin as an SLRP is involved with activities such as cell growth, proliferation and migration. It is an endogenous matrix-centric RTK inhibitor expressed by a target-specific tumour cell. This function may allow integration of decorin across multiple RTKs with differential binding kinetics for sustained and proficient cross talk towards tumorigenic suppression. In vivo experiments in male Wistar rats show decorin has a cardioprotective effect in type 2 diabetes associated with TGF-β1 (ERK1/2 and smad-2) and NF-κB pathways, decreasing the activation of IGF-IR while increasing expressions of inflammatory mediators such as PKC-α and Hsp70.

2.1.2 Biglycan

Biglycan plays a pivotal role in cell signalling, cell growth, cell differentiation and organising the assembly of collagen-1 fibres [27]. Despite its structural homology with decorin, biglycan does not present the ability of SLRP decorin to bind with the epidermal growth factor receptor (EGFR) in tumour cell regulation. Still, biglycan is considered to be a potential marker for different types of cancers associated with poor clinical outcome. In solid cancer/solid tumour, such as ovarian endometriosis, increased biglycan levels were reported to be isolated from serum, tissues and peritoneal fluid of patients in comparison to the control group [28]. Similarly, in *prostate cancer* biglycan overexpression significantly associates with only aggressive tumour phenotype that shows advanced pathogenesis of high Gleason grade, where grade 1 cells resemble normal prostate tissue and grade 5 cells are considered to be highly mutated barely resembling the normal cells that show accelerated cell proliferation along with lymph node metastases [29].

In pancreatic cancer biglycan is overexpressed, possibly due to the paracrine action of transforming growth factor beta (TGF-β) secreted by the stromal fibroblast cells [30]. In contrast, in vitro studies also show exogenous biglycan can inhibit the growth of pancreatic cancer cells by inducing cell cycle arrest and growth inhibitory function of the extracellular matrix [30, 31]. This finding is similar to the anti-proliferative properties as shown by decorin of the class I SLRP family.

Biglycan expression in *gastric cancer tissues* is associated with lymph node metastasis, depth of tumour invasion and tumour node metastasis (TNM) stage [32] inducing migration and invasion, allowing tube formation by the vascular endothelial cells in vitro [33]; animal experiment analysis done in vivo is consistent with the outcomes as investigated by in vitro investigations. Biglycan also increases phosphorylation of focal adhesion kinases (FAK), a non-receptor tyrosine kinase that participates in mechanically and chemically mediated mechanotransduction signalling events, and in gastric and pancreatic cancer FAK phosphorylates at Tyr576/577, Tyr925 and Tyr397 sites creating a binding domain for the src oncogene along with paxillin, a multidomain adaptor protein that occurs at the interface between the plasma membrane and actin cytoskeleton where it associates with focal adhesion kinases.

These findings show biglycan is upregulated and plays an oncogenic role in gastric cancer metastasis by activating the FAK signalling pathway [32].

In human *colon cancer*, biglycan overexpression upregulates VEGF level promoting angiogenesis and colon tumour growth by increasing cell viability that regulates tumour size, weight and CD-34 expression [34]. Extracellular signal-regulated kinase (ERK) pathway activates biglycan overexpression in colon cancer cells, and ERK inhibitor PD98059 dramatically reverses the increased expression of VEGF-induced biglycan [34, 35].

Thus, overexpressed BGN associates with the aggressive growth and metastasis of the tumours, with worse prognosis in patients with gastric and pancreatic carcinoma [32, 36]. These investigations suggest that biglycan may act as either a potential therapeutic target or prognostic biomarker in multiple types of cancer.

2.1.3 Asporin

Asporin (ASPN) also known as periodontal ligament-associated protein-1 (PLAP-1), is essential for the regulation of transforming growth factor beta-1 [37]. Asporin plays vital roles in the pathogenesis of different types of cancer and may act as an oncogene in pancreatic, colorectal, gastric and prostate cancer, as well as some types of breast cancer; however, asporin also acts as a tumour suppressor gene in triple-negative breast cancer.

In *pancreatic cancer* asporin is expressed by activated pancreatic stellate cells (PSCs) that allow the invasion and migration of pancreatic cancer cells (PCCs) [38]. Exogenous asporin interacts with the transmembrane receptor CD-44 on pancreatic cancer cells to activate nuclear factor kappa-B (NF-κB/p65)-mediated epithelial–mesenchymal transition (EMT) of cells involving protein kinase B (AKT), and extracellular signal-regulated kinase (ERK) pathways, including CD-44-induced NF-κB/p65 activation of pancreatic cancer cells in autocrine and paracrine signalling mechanisms; thus, asporin may be considered as a prognostic marker for targeting the tumour microenvironment in pancreatic cancer [39]. In human *colorectal cancer* (CRC) elevated expression of asporin occurs when compared to matched control with normal tissues [40], and 25% of these cancer cases show variation in the asporin gene structure [41]. Asporin promotes proliferation, migration and invasion of colorectal cancer cells by inhibiting apoptosis leading to the activation of Akt/Erk and TGF-β/Smad2/3 signalling [42]. Asporin interacts with Smad2/3 facilitating translocation of the peptides into the cell nucleus that can upregulate epithelial–mesenchymal transition-related genes. TGF-β signalling pathway is an essential regulator of EMT, as this pathway downregulates tight junction proteins resulting in weakened links, which is considered to be a key factor for EMT induction of TGF-β in colorectal cancer cell migration and invasion. Thus, high asporin expressions with variation in asporin gene structure indicate unfavourable prognosis, and is suggested to possibly serve as a potential biomarker in patients with colorectal cancer [41, 42].

Asporin is overexpressed in *gastric carcinoma* tissues compared to the corresponding noncancerous tissues, and their levels varies in gastric cancer epithelial cell lines [43]. Additionally, transient siRNA-mediated asporin knockdown in gastric cancer cells inhibits proliferation and suppresses the migration of immortalised neoplastic epithelial cells. Downregulation of asporin blocks the antiapoptotic molecule (Bcl-2) and increases the proapoptotic (Bad) molecule decreasing the levels of migration-related proteins (CD-44) and matrix metalloproteinase (MMP)-2 while abrogating the activation of ERK phosphorylation that involves expression of epidermal growth factors (EGF) and its receptors EGFR [44]. Asporin thus plays an oncogenic role in gastric cancer metastasis by influencing the EGFR signalling pathway.

Elevated expression of asporin is localised in *prostatic tumour* stroma that correlates with biochemical recurrence, progression, Gleason score and the clinical stages. The presence of asporin indicates it may be a stromal marker expressed by cells particularly with aggressive tumour subtypes [45]. In a mouse model of prostate cancer, asporin expression induced by p53 inhibition correlates with neuroendocrine markers. Asporin transcript in normal and cancer fibroblast cells is shown to be stimulated by conditioned media derived from the prostate cancer cells, but this is not reported in LNCaP, an androgen-sensitive, less metastatic prostate cancer cell lines [46].

Asporin acts as tumour suppressor in *breast cancer*, and high asporin expression significantly associates with better relapse-free survival in low-grade tumour patients becoming worse as the tumour increases to grade 3 stages [47]. Asporin expression in breast cancer associates with fibroblast cells as shown by in situ hybridisation technique incluidng Hs578T breast cancer cell line as evidenced by quantitative RT-PCR and western blotting. Asporin expression is downregulated by bone morphogenetic protein-4, and shRNA inhibits Hs578T, CAFs and T47D breast cancer cell line, where as upregulation may be facilitated by serum-free three-dimensional growth in stiff Alvetex scaffold. Molecular peptides/fragments of asporin are also involved in the invasion of breast cancer cells, and lack of asporin expression is related with poor prognosis in breast cancer [48].

2.1.4 Lumican

Lumican is ubiquitously occurring SLRP molecule present in the normal extracellular matrix, and plays a role in controlling cell growth, invasion and attachment. In cancer cells, lumican acts by regulating cellular expression that increases the availability of growth factors in the stromal tissue while also inhibiting tumour progression that varies according to the organs, stages and type of tumours [49]. In *lung cancer*, including adenocarcinoma and squamous cell carcinoma, lumican expression is upregulated, and immunohistochemical analysis shows them to be localised in the cytoplasm of the cancer cells and/or stromal tissues adjacent to the malignant cells [50]. Lumican in adenocarcinoma of lungs correlates with pleural invasion and large size tumour formation in comparison to the stromal tissue with similar clinicopathological features [51], while in lungs squamous cell carcinoma with a peculiar keratinisation pattern, the stromal cell expresses lumican that correlates with vascular invasions [52]. However, in both adenocarcinoma and squamous cell carcinoma, lumican expression does not correlate with patient prognosis. Further, lumican mRNA is expressed in LC-1/Sq and EBC-1 squamous cell carcinoma line, and A549, RERF-LC-KJ and PC-3 adenocarcinoma cell lines [53]. Moreover, the molecular size of lumican extracted from the lung cancer tissue differs from that in the cell culture lines possibly due to the presence of glycosylated side chains that are attached to the core protein molecule [54]. These findings indicate that glycosylation of lumican within cancer cells and the stromal tissues may correlate with the aggressiveness of lung squamous cell carcinoma and adenocarcinoma. In normal lung tissues, lumican is therefore weakly localised in the vascular smooth muscle cells, perivascular, peribronchial connective tissues and bronchial epithelium [55].

Lumican in *pancreatic cancer* tissue is mainly localised in the acinar and islet cells, and in chronic pancreatitis-like lesions lumican expression occurs adjacent to the tumour tissue, and even in the fibroblast cells with dense collagen fibres that are close to the cancer cells, lumican thus associates with poor prognosis in human pancreatic cancer [56]. An investigation by immunocytochemical analysis in pancreatic cancer shows the tumour cells tend to be devoid of either decorin, lumican or versican, and further RT-PCR analysis confirms pancreatic stellate cells as the major source of these proteins [57]. A large 70-kDa lumican core protein stimulates growth but inhibits the invasion of human pancreatic cancer cells, and culture medium expressing lumican applied to PANC-1adenocarcinoma cell line stimulates growth but inhibites cell invasions [56]. The 70-kDa lumican activates extracellular signal-regulated kinase (ERK) and induces adherence to laminin and alpha-3 integrin subunit decreasing the active MMP-9 enzyme in PANC-1 cell line; thus, it has been suggested that secreted 70-kDa lumican may be involved in pancreatic ductal adenocarcinoma cell growth, invasion and attachments.

In *colorectal cancer*, lumican is strongly expressed in 75% of the cases analysed and spatially localised within epithelial, fibroblast and colorectal cancer cells with mild reactive dysplasia [58], suggesting lumican synthesised in colorectal cancer tissue may develop dysplasia. Overexpression of lumican affects the migration of human colon cancer cells through upregulation of gelsolin, a calcium-dependent actin binding protein that reorganises cell migration events in cancer [59]. Further, a study on rectal carcinoma analysed the sensitivity of the cancer cells in response to preoperative radiotherapy by DNA microarray gene expression profiles, and the results showed apoptosis inducers (lumican, thrombospondin 2 and galectin-1) were highly expressed in responders, whereas apoptosis inhibitors (cyclophilin 40 and glutathione peroxidase) were higher in nonresponders; thus, gene expression profiles in response to radiotherapy may be useful in predicting outcome responses for individualised therapy in rectal cancer [60].

Analysis on the role of lumican in *prostate cancer* pathogenesis by in vivo and in vitro investigation shows lumican upregulates in the primary tumour cells and the reactive stroma surrounding the tumours with fibrotic deposition circling the acinar glands [61], and in vitro analysis demonstrates lumican inhibits migration and invasion of metastatic prostate cancer cells isolated from the lymph node, bone and brain tissue. Moreover, prostate cancer cells seeded on lumican scaffold decrease the formation of cell projections lamellipodia detected as rearrangement in keratin-8/18, integrin-β1, membrane type 1 matrix metalloproteinases (MT1-MMP) and invadopodia following disruption of α-smooth muscle actin, cortactin and neural Wiskott-Aldrich syndrome protein (N-WASP) [62]. In lumican knockout mice, a significant increase in prostate cancer cell invasion was observed to occur through the peritoneum lining the abdominal cavity, suggesting the restrictive role of lumican in the extracellular matrix during prostate cancer invasion. Lumican present in the reactive stroma surrounding prostatic primary tumours can also be a reason for the restrictive prostate cancer progression, and thus it has been postulated that lumican could act as a biomarker in prostate cancer staging [63].

In *breast cancer*, lumican may inhibit or even reverse several metastatic features through oestrogen receptors ERα and ERβ that is associated with epithelial-to-mesenchymal transition (EMT), a significant factor that permits breast cancer cell proliferation, differentiation and progression [64]. Oestrogen receptors aid the link between cell–cell and cell–matrix; therefore, the absence or presence of these receptors plays a vital role in breast cancer cell progression [65]. The endocrine resistance in epithelial cells increases cell proliferation, resulting in a mesenchymal phenotype facilitating epithelial-to-mesenchymal transition affecting transcriptional factors that governs the transition in breast cancer cells. Knockdown or silencing of endocrine resistance in MCF-7 and MDA-MB-231 breast cancer cell line provokes pivotal changes in cell phenotype, functions, mRNA expression and protein levels of EMT markers [66]. Mesenchymal cells show migratory and invasive growth by invadopodia, while in the epithelial cells, lamellipodia and filopodia are shown to be involved. Invadopodia are actin-rich protrusions of plasma membrane that promotes proteolytic degradation of the matrix in tumour invasions [67]. Cortactin and MMP-14 are also suggested to be essential in the formation and functions of invadopodia. In vitro experiments show lumican inhibits cortactin and MMP-14 expression, altering the formation of lamellipodia, and transforms mesenchymal cells into epithelial-like cells [68, 69]. Therefore, a lumican-based peptide with anti-breast cancer properties may pharmacologically target to inhibit EMT.

Osteosarcoma is a common type of primary bone tumour in children and adolescence between 10 and 19 years of age [70], and lumican expression positively correlates with cell differentiation but negatively with osteosarcoma progression in moderately differentiated high metastatic and well-differentiated low metastatic human osteosarcoma cell line [71]. The human osteosarcoma cell-secreted lumican can be substituted with a keratan sulphate glycosaminoglycan side chain. Low metastatic, well-differentiated (Saos-2) cells express lumican sevenfold higher than the high metastatic MG-63 cells. Interfering RNA to lumican gene downregulates their expression in both of these cell lines, where growth of Saos-2 cells is inhibited, but the migration and chemotactic responses to fibronectin molecule are promoted. Lumican further negatively correlates with Smad-2 activation indicating the viability of phosphorylated TGF-β2 receptors in osteosarcoma [72]. In Saos-2 silenced RNA and scramble siRNA cells, adherence of fibronectin to lumican and neutralisation with endogenous transforming growth factor-β2 (TGF-β2) do not produce any difference. However, exogenous TGF-β2 stimulation of Saos-2 cells promotes adhesion to fibronectin-coated surfaces, resulting in cell adhesion, and thus shows dependence on active TGF-β2 signalling. Furthermore, significant increase in Smad-2 activation occurs in lumican silenced cells in the presence of anti-TGF-β2 peptide demonstrating lumican can act as upstream regulator of TGF-β2/Smad-2 signalling cascade [73]. Further, fibronectin-dependent adhesion, β1-integrin expression and phosphorylated foal adhesion kinase activation tend to downregulate TGF-β2-regulated lumican expression [74]. In contrast, highly aggressive MG-63 human osteosarcoma cell line is nonsensitive to decrease in endogenous level of lumican.

Melanoma is a type of skin cancer that occurs due to the ultraviolet exposure of melanin-producing skin cells that may occasionally spread to other parts of the body [75]. Lumican shows antitumour properties in melanoma, and a 37-kDa low-molecular-mass lumican core protein increases melanoma cell adhesion, while a 57-kDa large, glycosylated protein core inhibits cell migration and invasion [55, 76]. Furthermore, a 17-amino-acid sequence within nine LRRs is able to produce anti-migratory effect of lumican in vitro which is localised to stromal dermal fibroblast cells at the periphery of the tumour site [77]. Lumican expression is shown to depend on the melanoma cell line used, and WM9 and M5 cells secrete them as a core protein with glycosylated keratan sulphate side chain, while in Snail-B16F1 cell line lumican inhibits anchorage-independent cell proliferation, migration, invasion and MMP-14 activity [78]. In an in vivo mouse model, B16F1 melanoma cell line transfected with the human lumican cDNA significantly inhibits tumour progression by downregulating p21 pathway, and lumican demonstrates to inhibit melanoma cell migration by altering cytoskeletal actin network, focal adhesion kinase complexes and integrin alpha-2 beta-1 receptor expression to which lumican binds directly [79, 80]. The lumican protein of 57-kDa molecular mass has angiostatic properties that inhibits endothelial cell invasion, angiogenic sprouting and vessel formation in mice and is associated with the activity of MMP-9 and MMP-14 via the integrin receptors [81].

2.1.5 Fibromodulin

Fibromodulin (FMOD) assists with collagen fibrillogenesis, a process involved in the assembly of collagen fibrils necessary for matrix organisation and tissue repair of many organs. It also regulates angiogenesis, TGF-β modulation, differentiation of human fibroblast cells, inflammation, apoptosis and cellular metastasis [82].

Fibromodulin acts as tumour-specific antigen in leukaemia, leiomyoma and glioblastoma, and in cancer FMOD increases cell migration and angiogenesis by suppressing apoptosis, and via its interaction with nuclear factor kappa-B (NF-κB) pathway to progress as malignancy [83]. FMOD is thus suggested to be of diagnostic and therapeutic applications in cancer therapy.

Chronic lymphocytic leukaemia is a cancer of the blood and bone marrow commonly affecting the older adults above the age of 65 years [84]. Fibromodulin gene expression positively associates with B-cell chronic lymphocytic leukaemia (B-CLL) patients and negatively with age- and sex-matched control healthy volunteers as none of them shows peripheral blood mononuclear cells positivity to fibromodulin gene expression [85]. High fibromodulin mRNA expression in B-CLL patients also significantly associates with risk factors such as hepatomegaly, lower haemoglobin level, lower RBC count, lower platelet count, lymphadenopathy and splenomegaly [86]. At the protein level, fibromodulin is localised in the cytoplasm of the B-CLL cells and in the supernatant after in vitro cell culture analysis; however, they are not detected in cell surface. Fibromodulin is also characterised as tumour-associated antigen (TAA) in B-cell chronic lymphocytic leukaemia (B-CLL) with the potential to elicit specific antitumor responses [87]. This may be possibly due to the CLL cell phenotypes downregulating proapoptotic genes, together with upregulation of the antiapoptotic molecules (Bcl-2). CLL cells overexpress EPAC and CDC25 genes encoding guanine nucleotide exchange factors that activate Rap-1 and Ras, which are small GTPases that control pleiotropic transcriptional responses via the Raf/ERK pathway [88]. Deregulation of Ras and Rap-1 pathway represents one of the most common alterations in various human tumours. CLL upregulates mRNA expression of several cytokine or chemokine receptors such as IL-4R, TGF-β type III receptor and CCR7, and this suggests that CLL cells may be abnormally responsive to certain stimuli that may potentially involve the matrix proteins including SLRPs [89].

Glioblastoma multiforme (*GBM*), also called as astrocytoma, is a fast-growing, aggressive form of intracranial brain tumour that generally does not spread to distant organs. It is more common in males in comparison to females with a median diagnostic age of 64 years and a survival time of 14 months [90]. Integrated genomics involving DNA methylation and transcriptome profile analysis identifies fibromodulin as a glioblastoma upregulated gene [91], and secreted fibromodulin tends to promote glioma cell migration through the formation of filamentous actin stress fibres [83]. Small interfering RNA and inhibitor of actin polymerisation such as cytochalasin-D show fibromodulin-induced glioma cell migration is dependent on integrin-FAK-Src-Rho-ROCK signalling pathway [92]. Further, the C-terminus of fibromodulin is less conserved and undergoes fragmentation readily at the leucine-rich repeat domain, thus promoting nonbinding to collagen type I which consequently fails to induce integrin signalling and promote glioma cell migration. Fibromodulin-induced integrin activation and migration can be abrogated by a 9-mer wild-type peptide at the C-terminus site, and the same peptide with mutation in LRR residues is suggested to be essential for interaction of fibromodulin with collagen type I involving the integrins [91]. Chromatin immunoprecipitation analysis shows transforming growth factor beta-1 (TGF-β1) regulates fibromodulin expression through epigenetic remodelling of demethylation promoter, gain of active histone with simultaneous loss of DNA methyltransferase-3-alpha (DNMT3A) and as an enhancer of zeste homolog-2 (EZH2), followed by enrichment of Smad-2 Mad-related protein-2 (SMAD2) and paralogous transcription of coactivators as stimulus-dependent factors [93]. Fibromodulin silencing thus inhibits TGF-β1-mediated glioma cell migration where secreted FMOD and its peptides act as an important regulator of glioma cell migration and is therefore a potent therapeutic intervention in glioblastoma.

Prostate cancer incidence increases with age and genetic alterations, and is the second most frequent type of neoplasm in males [94]. Fibromodulin gene expression increases during prostate cancer, where the FMOD gene is involved in collagen fibrillogenesis, cell adhesion and modulation of TGF-β signalling pathways. FMOD gene sequencing and analysis of the coding region in prostate cancer show six variants, of which two are missense alterations at p.(Tyr42Ser) and p.(Pro24Ala), three are synonymous at p.(His253=), p.(Asn353=) and p.(Glu79=) and one of them is intronic (c.980-114A > G). The p.(Tyr42Ser), p.(Pro24Ala) and p.(Asn353=) are rare gene variants, while p.(Tyr42Ser) is a potential pathological alteration; these variants are not observed in the control non-prostatic cancer group, suggesting that fibromodulin gene may be a potential biomarker in the development of prostate cancer [95].

Gene transcription and expression analysis in the prostate cancer cell lines LNCaP and PC-3 further shows metastatic phenotypes involving caveolin-1 (CAV-1), prostate cancer marker (CD-24), endothelial cell-specific molecule-1 (ESM-1) and trypsin receptor-1 (F2RL-1) overexpressed by the aggressive PC-3 cell line, and in less aggressive LNCaP cells CAV-1 and ESM-1 are undetected [96]. The encoding protein caveolin-1 is involved in molecular trafficking, cell adhesion and signal transduction mechanism; thus, associates with the development of prostate cancer, and it is also detectable in blood at a significant level which could assist in diagnosis and prognosis of prostate cancer [97]. CD-24, a cancer stem cell surface marker, encodes glycosyl phosphatidylinositol-linked sialoglycoprotein, which is expressed by prostate cancer cells, and overexpression of CD-24 acts as a predictor of relapse and poor outcome prognosis in various cancer types [98]. Endothelial cell-specific molecule 1 (ESM-1), also known as endocan, regulates adhesion of circulating cancer cells to the endothelial vasculature facilitating invasions [99]. ESM-1 thus interacts with angiogenic growth factors enhancing metastatic abilities and may alternatively monitor tumour response to antiangiogenic therapy as observed in renal cancer. F2RL-1 is a novel biomarker gene overexpressed in PC-3 cell line; this gene is associated with new blood vessel formation in retinal development and ischemic retinopathy, and further evaluation of FMOD gene transcript in patients diagnosed with benign or malignant prostatic disease identifies overexpression of these gene; however, their specific mechanism of action still remains underexplored [100].

Leiomyomas, also called fibroids, is the most common benign tumours of the uterus in premenopausal women, arising from the uterine myometrium including smooth muscles and connective tissue fibroblast cells [101]. A characteristic feature of leiomyomas is structural disorientation and accumulation of the extracellular matrix molecules around the tumour region that increases tissue stiffness and contributes towards modified signalling events, altered mechanotransduction activities and tumorigenesis [102]. Past studies show several specific gene mutations occur in leiomyomas, and microarray gene profiling reveals fibromodulin is differentially expressed and significantly upregulated in leiomyomas, specifically two- to threefold of higher expressions is reported, in comparison to uterine myometrium in the proliferative phase [103]. Further, western blotting and immunohistochemical analysis show detectable fibromodulin level in leiomyomas from the myometrial tissue extract, leiomyoma smooth muscle cells (LSMC), myometrial smooth muscle cells (MSMC), connective tissue fibroblasts and the arterial walls. Time-dependent cell culture analysis shows TGF-beta increases fibromodulin expression in MSMC, and the gonadotropin-releasing hormone analogue (GnRHa) inhibits fibromodulin expression in MSMC and LSMC. Additionally, the effect of TGF-beta and GnRHa on fibromodulin expression can be reversed following pretreatment of LSMC and MSMC with Smad3-siRNA, an inhibitor of TGF-beta inducible promoter activity, that also involves U0126, a MEK1/2 signalling pathway inhibiting cell growth and differentiation [104]. These results therefore indicate fibromodulin is differentially regulated in leiomyomas.

Mechanical stress due to excessive extracellular matrix deposition is a critical factor in tumorigenesis, and uterus being a reproductive organ is distensible and viscoelastic. Thus, in leiomyomas, the cells and the corresponding matrix undergo significant changes in dynamic and static mechanical signalling due to alterations in strain and pseudo-dynamic stiffness modulus relative to the matched normal myometrium [105]. Moreover, leiomyoma cell in culture demonstrates an attenuated response to applied cyclic uniaxial strain and to variation in the anchoring substrate stiffness when compared to normal myometrial cells. Leiomyoma and myometrial cells when grown on a flexible pronectin-coated silicone substrate at basal levels and in lysophosphatidic acid-stimulated condition show similar RhoA activity, a regulatory factor that controls the cytoskeleton of the cells. In contrast, when leiomyoma cells are plated on a rigid polystyrene substrate they show elevated levels of active RhoA in comparision to the myometrial cells. Thus, leiomyomas contribute to inherent tumour tissue stiffness, and exhibit sensitivity to mechanical signals through RhoA activity that upregulates cytoskeletal protein involved in cell migration [106]; however, further research is required to elucidate the specific mechanisms responsible in the mechanical signalling of cells in leiomyomas.

Small cell lung cancer (SCLC) is a rare and aggressive form of carcinoma of the respiratory system and remains relatively undiagnosed until the advanced stages. Although small cell lung cancer is sensitive to radiotherapy and chemotherapy, it is highly prone to recurrence resulting in poor prognosis [107]. Small cell lung cancer is an angiogenesis-dependent disease that allows the formation of new blood vessels from the pre-existing endothelial cells, facilitating the release of growth factors and cytokines that stimulates the tumour cells and the extracellular matrix [108]. Fibromodulin is shown to promote angiogenesis in cutaneous diseases, wound healing and optical (eye) tissue, while being significantly expressed in small cell lung cancer tissue at late stage where the cancer cells migrate out of the effected lung to the opposite side or even distant organs with higher vascular density [109]. Animal studies further show that xenograft tumours developed with fibromodulin silenced H446 cells (H446-shFMOD) exhibit slowed growth rate, decreased microvessel density and reduced blood perfusion in comparison to controls [110].

H446-shFMOD culture cells also show decreased secretion of fibromodulin in conditioned medium that could inhibit proliferation, migration and invasion of human umbilical vessels (comprising of endothelial cells) compared to controls. Decreased secretion of fibromodulin downregulates the expression of vascular endothelial growth factor (VEGF), transforming growth factor beta-1 (TGF-β1), fibroblast growth factor-2 (FGF-2) and platelet-derived growth factor-B (PDGF-B) in human umbilical endothelial cells [111]. Fibromodulin expressed by small cell lung cancer may promote tumour angiogenesis, and silencing of fibromodulin has been suggested as a potential clinical therapy for repressing tumour angiogenesis.

Experimental carcinoma is a treatment-based approach for identifying effective drugs and target bioregulatory markers in solid malignancies involving the tumour stroma [112]. Invasiveness to promote growth of malignant cells, abnormal blood vessels and connective tissue cells that produce a fibrotic collagen-rich matrix characterise the stroma of experimental carcinoma. A mathematical model of cancer invasion suggests heterogeneous extracellular matrix with altered matrix components stimulate aggressive tumour cell phenotypes, whereas tumours with a homogenous matrix are less aggressive and invasive [113]. Fibromodulin mRNA has been detected in a variety of clinical malignancies, such as leiomyomas, breast and prostate cancer. Thus, to elucidate the effect of fibromodulin expression in experimental carcinomas, animal model, xenografts and cell lines are explored, including chemically induced rat mammary gland carcinoma, mouse RIP-Tag insulinoma, transplantable syngeneic colonic carcinoma and KAT-4 carcinoma grown as a xenograft in mice and numerous tumour-specific cell lines, where fibromodulin promotes the formation of a dense collagen scaffold and illicit inflammatory responses to promote elevated synthesis of interstitial fluid pressure [114]. These changes are conducive to the formation of distorted blood vessels, and fibromodulin tends to control the structure of tissue as well as the fluid balance in experimental carcinoma [115]. Additionally, fibromodulin knockdown model does not affect the blood vasculature structure but increases the extracellular fluid volume and lowers the interstitial fluid pressure. Thus, fibromodulin controls the stroma of the extracellular matrix and modulates fluid convection inside and out of the stroma. These findings are suggested to be particularly important in relation to the fluid balance modifying response in cancer therapeutics [83].

Breast cancer is one of the most common cancer types in women with no specific risk factor other than gender and age [116]. Breast cancer arises due to abnormal growth of cells inside the lactiferous (milk) ducts or lobules of the breast which can spread into the surrounding breast tissue by invasion [117]. The early stages of breast cancer are not life-threatening and can be detected by mammogram, while late-stage metastasis is invasive affecting the nearby lymph nodes or other organs and can be life-threatening [118]. Breast cancer treatment is based on stages and may combine surgery, radiation therapy and medications. Invasive breast cancer shows reduced expression of two SLRPs, and they are fibromodulin and decorin [119]. In vitro and in vivo analysis shows fibromodulin is a potent TGF-β-antagonist, when compared to decorin inhibiting neointimal hyperplasia in saphenous vein graft. TGF-β plays a role in the regulation of cell growth, differentiation and migration, and an association between allele variants of TGF-β1 and invasive breast cancer has been reported. Fibromodulin contributes towards significant reduction in TGF-β1 expression levels in the highly metastatic 4 T1 breast cancer cell line [120]. Inhibition of fibromodulin expression in animal model leads to delayed wound healing and increased scar size formation, and overexpression decreases TGF-β1 levels [121]. Fibromodulin also shows inhibitory effect to NF-κB signalling through suppression of inhibitors of nuclear factor kappa-B (IκBα) in 3 T3-L1 fibroblast cell line [122]. NF-κB signalling stimulates proliferation, migration, invasion including angiogenesis and metastasis while also inhibiting apoptosis in breast cancer.

Similarly, recent investigations also show fibromodulin in breast cancer regulates cell migration and invasion via activation of ERK, and the Wnt/β-catenin pathway, where the β-catenin/TCF4/LEF1 promoter transcribes fibromodulin [123]. Aspirin is suggested to inhibit cell migration and invasion by attenuating Wnt/β-catenin signalling suppressing fibromodulin expression by deacetylation of β-catenin histone deacetylase-6 (HDAC6) leading to phosphorylation and cytoplasmic degradation [124]. These findings expand the understanding of Wnt/β-catenin pathway in tumour metastasis and reveals a potential mechanism regulating the anticancer activity of aspirin, which may be valuable in the development of cancer therapeutics.

Colorectal cancer is the third most commonly occurring cancer of the colon and rectum and accounts for approximately 10% of all the reported cancer cases worldwide [125]. It is also the second leading cause of death and affects both sexes but predominates in older individuals aged 50 years and above. Several risk factors contribute to the development of colorectal cancer such as age, family history of colorectal cancer, high intake of processed meats, low consumption of fruits and vegetables, sedentary lifestyle, obesity, smoking and excessive alcohol intake [126]. Early detection of colorectal cancer leads to better treatment outcome, late stages are often fatal and intermediate stage treatment strategies depend on progression, spread of the cancer and the patient medical history [127].

Fibromodulin is significantly elevated in human colorectal cancer cell lines and includes HCT116, LoVo and $Caco_2$-derived epithelial cells of the patients with colon cancer [128]. Further, it has been also shown that human colorectal cancer tissue shows higher levels of fibromodulin than the adjacent stromal tissue (more than 2 cm away from the tumour edge) [129].

Fibromodulin in colorectal cancer can regulate the immune microenvironment and negatively correlates with $CD8^+$ T cells and NKT cells that modulate immune response against infectious stimuli, antigens, tumours and tissue grafts [130].

According to single-cell RNA sequencing data, epithelial and stromal cells have the highest expression levels of fibromodulin in colorectal cancer [131]. In vitro and in vivo, screening of 12-mer peptide RP4 targets fibromodulin expression thorough five rounds of biopanning shows specific blocking activity, where RP4 directly inhibits the growth and metastasis of colorectal cancer cell lines HCT116, LoVo and CT26, by downregulating the AKT and Wnt/-catenin signalling pathways [130]. Thus, fibromodulin expression is clinically significant with colorectal cancer progression, and absence may promote tumour growth by regulating tumour-associated immune cells [132].

2.1.6 PRELP

PRELP (proline/arginine-rich end leucine-rich repeat protein) also called as prolargin is abundant in collagen-rich tissues, and they even bind collagen to basement membranes of epithelial cells and cartilage [133]. The role of PRELP in various types of cancer is still emerging, and recent evidence suggests that non-glycosylated form of PRLEP fragment with a 38-kDa size which is smaller than the normal 50-kDa core protein is associated with various types of carcinomas [134]. Overexpression and underexpression of PRLEP is associated with poor patient prognosis and promotes cell migration and proliferation of hepatocellular carcinoma (HCC) cell lines [135]. The specific role of PRLEP in various types of cancer is described in the following subsections.

Hepatocellular carcinoma (HCC) is the most common type of primary liver cancer that occurs in patients with chronic liver diseases caused by hepatitis B or C virus infection, alcohol consumption and accumulation of fats in the liver [136]. Treatment measure includes HBV vaccination, surgery including liver transplant, chemotherapy or radiation, drugs and immunotherapy, but these treatments are effective only in the early stages as late stage is often untreatable, and surgeries or transplant cannot be recommended due to metastasis and lacks success in controlling the spread of disease.

Tumour proliferation and metastasis are the significant hallmark of hepatocellular carcinoma [137], and PRLEP knockdown promotes tumorigenesis, while overexpression inhibits cell proliferation and migration. PRELP is generally downregulated in hepatocellular carcinoma tissues in comparison to patients with matched hepatic cirrhosis. PRELP expression is also found to be significantly associated with patient prognosis, and low expression tends to indicate shorter overall survival time and relapse-free survival time in patients [135, 138]. Furthermore, low tissue expression levels of PRELP are expressed in large tumour and advanced pathological stage. PRELP is thus downregulated in hepatocellular carcinoma and has been suggested to act as a potential prognostic predictor.

Bladder cancer affects the urinary system and arises from the cells and the tissues that line the urinary bladder, a hollow reservoir for storing urine. Urothelial carcinoma is the most common form of bladder cancer and prevalent among older adults [139]. The risk factors include smoking, chemical exposure, chronic bladder inflammation and genetic susceptibility. Common symptoms are gross or microscopic haematuria, urinary frequency and pelvic pain. Bladder cancer is classified by stages in muscle invasion of cancer cells and also by grades as low or highly metastatic. Based on the morphology, it may also be subdivided into papillary, low malignant potential papillary carcinoma and urothelial carcinoma in invasive categories [140]. Treatment includes transurethral resection of bladder tumour, intravesicle chemotherapy, immunotherapy and therapeutic drugs [141].

Proline/arginine-rich end leucine repeat protein (PRELP) are specifically expressed by the umbrella-shaped cells lining the epithelium of the urinary bladder, and PRELP levels are significantly downregulated in bladder cancers from early stages involving various types of epithelial cells [142]. In vitro studies including gene expression profiling using bladder cancer cell lines revealed PRELP application suppresses the cancer progression by inhibiting TGF-β and EGF pathways, which reverses epithelial–mesenchymal transition (EMT), activates cell–cell adhesion and inhibits various oncogenic pathways [143]. Furthermore, in the bladder epithelia, knockout mice model of PRELP gene causes partial EMT and loss of tight junctions of the umbrella-shaped cells, resulting in the formation of bladder carcinoma (in situ) by spontaneous breakdown of the umbrella shaped cell layer. Data indicate PRELP are endogenous inhibitors of cancer initiation and progression as they control EMT and may potentially be used towards the treatment of bladder cancer [142].

Further, recent findings also suggest that pleomorphic adenoma gene-like 2 (PLAGL-2), a proto-oncogene, functions as a crucial regulator of epithelial–mesenchymal transition (EMT) factor to promote bladder cancer cell invasion and metastasis. Upregulated PLAGL-2 is linked to unfavourable survival, and PLAGL-2 deficiency significantly inhibits the proliferation and metastasis of bladder cancer cells [144]. Furthermore, PLAGL-2 inhibits the Hippo pathway that facilitates the nuclear translocation of YAP1/TAZ, a transcription coregulator involved in cellular proliferation and apoptosis suppression, increasing YAP1/TAZ activity mediated by RACGAP1-RhoA activation. Therefore, PLAGL-2/RACGAP1/RhoA GTPase/YAP1 signalling is a potential pathway for bladder cancer therapy.

Retinoblastoma (RB) is a rare intraocular paediatric cancer which develops before the age of 5 years [145]. Most cases of retinoblastoma are due to non-heritable causes but involves mutation in the RB1 gene [146]. Heritable retinoblastoma increases the child risk of acquiring other forms of cancers such as lungs, bladder cancer and tumours of the brain and pineal gland [147]. Genetic test focusing on the RB1 gene can determine whether a child has heritable or non-heritable retinoblastoma, and treatment plan includes chemotherapy, thermotherapy, radiation and surgical excision which is unfortunately accompanied by side effects such as vision and hearing loss [148].

Human retinoblastoma whole genome sequencing and mRNA profiling followed by subsequent analysis of retinoblastoma cell line Y29 and WERI-RB1confirms PRLEP is suppressed in the retina of the eyes and RB1 gene mutation is the direct consequence behind PRELP suppression [149]. Similarly, in mouse model with PRELP inactivation leads to dysplasia of the retinal cell layers, especially the inner nuclear layer, outer nuclear layer and the ganglionic cell layer [150], and PRELP suppression contributes towards weakening of the retinal cell adhesion facilitating epithelial mesenchymal transition (EMT) enhancing the progression of retinoblastoma. PRELP directly binds to multiple extracellular protein receptors such as TGF-β, EGF, IGFI and p75NTR components that are known to regulate cell cycle, EMT, cell adhesion and inflammation [151]. Furthermore, application of PRELP to retinoblastoma cell lines negatively regulates oncogenic pathways such as RAS, NF-κB and AKT, followed by cell cycle inhibition and activation of apoptosis that impacts cell–cell adhesion properties [149]. These observations indicate PRELP plays a fundamental role in maintaining retinal microenvironment and may act as a novel tumour suppressor against retinoblastoma.

Pancreatic ductal adenocarcinoma (PDAC) is the most common type of pancreatic cancer forming 80% of the reported cases and is the fourth most common cause of global cancer-related deaths [152]. The prognosis of PDAC is extremely poor, and majority of the patients die within 6 months of diagnosis, and only 6% of the patients survive for 5 years [153]. Risk factor includes a family history of the diseases, pancreatitis (inflammation of pancreas), diabetes, obesity and smoking. Treatment option may depend on the extent of the disease spread and include surgery, radiation, chemotherapy, chemoradiation and targeted therapy [154].

PRELP was reported to be one of the differential proteins associated with pancreatic ductal adenocarcinoma especially among the rare very long-term survival group of patients [153]. Based on the proteomic data, PRELP was found to be 3.3-fold more abundant in the tumour tissues in the very long-term survival group compared to the short-term survival patient group. PRELP protein in very long-term survival group patients showed a mass spectral count ranging from 9 to 23 in contrast to PRELP being marginally detectable in short-term survival group patients with mass spectral count ranging from 1 to 7. Immunohistochemical analysis performed on tissue sections further revealed neoplastic epithelial cells displayed varying degrees of PRELP staining, while peritumoral stromal cells were mostly negative, and weak immuno-ositivity was seen in occasional cases. No significant associations were noted between PRELP abundance and baseline clinicopathologic parameters including age, gender, tumour grade, tumour size, tumour stage, lymph node involvement and tumour status. Neoplastic cells of patients with longer survival time displayed higher intensity of PRELP protein staining than the tissues from patients with shorter survival time.

Colorectal cancer (CRC) is a common malignant tumour of the colon and rectum of the large intestine [155]. The incidence and mortality rate are second only to gastric cancer, oesophageal cancer and primary liver cancer of the digestive system. However, the cause of its development is not fully elucidated.

PRELP inhibits the development and progression of a variety of cancer as described above, and in human colorectal cancer PRELP expression is found to be significantly higher in metastatic tissues than in non-metastatic part of the colorectal cancer [156]. PRELP expression closely associates with poor prognosis of the disease and promotes growth and metastasis of cancer cells by reducing cell stiffness and allowing adhesion through epithelial mesenchymal transition (EMT). PRELP binds to integrin alpha-5 to activate the integrin α5/FAK/AKT signalling pathway, known to facilitate cell proliferation, migration and adhesion. PRELP is upregulated in metastatic colorectal cancer and thus can be used as a potential prognostic marker or therapeutic target for the clinical management of metastatic colorectal cancer [156].

Chronic lymphocytic leukaemia (CLL) or *small lymphocytic lymphoma (SLL)* is an indolent monoclonal lymphoproliferative malignancy characterised by increased production of mature but dysfunctional B lymphocytes [157]. The leukaemic phase is often referred to as chronic lymphocytic leukaemia, which can be restricted to lymph nodes as small lymphocytic lymphoma. The primary disease sites include peripheral blood, spleen, lymph nodes and bone marrow, but the main reason behind occurrence of CLL is still unknown, and genetic mutation, environmental factors, exposure to benzene and smoking are suggested to be the likely cause behind its occurrences [158]. Though CLL is more common in older adults with an average age of 70 years, it has been detected in younger aged adults of 30 years. Patients diagnosed with CLL have limited treatment plan as the disease is heterogeneous and uncurable, and present preventive strategies include allogenic hematopoietic stem cell transplantation, therapeutic drugs such as BCL-2, PI3K inhibitors and purine analogues which are some of the few options utilised in the management of CLL [159].

Overexpression of common genes in tumour cells occurs due to epigenetic regulations of closely located gene clusters, PRELP is structurally similar and located 80 kbs proximal to fibromodulin gene on chromosome 1q32.1 and fibromodulin is overexpressed in CLL [134]. Thus, PRLEP gene, cell lines and mRNA expression analysis in CLL show it is uniquely present along with fibromodulin. Other haematological malignancies including cell lines and tissue from normal donors do not express PRELP. Strong polyclonal activation of normal B and T lymphocytes does not induce PRELP expression, suggesting that CLL might be a constitutive tumour-related aberration in vivo. The expression of PRELP to age, gender, leukaemic cell count, genomic aberrations (13q del, 17p del, trisomy12) and previous treatment has no correlation. Further, 50- and 58-kDa PRELP proteins are detected in serum from CLL patients and healthy controls, which may correspond to mature glycosylated expression of PRELP possibly being produced by fibroblast cell in the connective tissues [160]. However, in the CLL cell lysates, a unique 38-kDa PRELP protein fragment is detected by both C-terminal and N-terminal anti-PRELP antibodies indicating that 38-kDa protein is not a cleavage or degradation product. This is further supported by deglycosylation experiments that reduced all PRELP bands with the exception for the 38-kDa band. The presence of an intact peptide suggests that 38-kDa PRELP may be retained in the cytosol, and thus they are not detected in serum or are secreted by CLL cells [134]. Mutation analysis of the PRELP gene in CLL does not reveal any substantial nucleotide aberrations, and these findings, in combination with only two coding exons, makes splice variants or truncation of PRLEP unlikely.

2.1.7 Keratocan and Pancreatic Cancer

Keratocan and pancreatic cancer is characterised by desmoplasia, a prominent pathological feature marked by increased proliferation of alpha-smooth muscle actin-positive fibroblasts also referred to as activated pancreatic stellate cells. Desmoplasia results in increased and altered deposition of extracellular matrix molecules that reduces tissue elasticity and increases the interstitial fluid pressure contributing to chemoresistance in pancreatic cancer [161, 162].

Keratocan along with the keratan sulphate proteoglycan side chain is shown to be associated with pancreatic cancer [163], and lumican, a class II SLRP, acts as a novel modulator of keratocan gene expression by keratocytes regulating collagen fibrillogenesis [164]. Lumican expression is associated with pancreatic cancer including several other biological processes such as desmoplasia, wound healing, epithelial–mesenchymal transition and tumorigenesis [165]. Therefore, evidence from gene expression profiling interactive analysis (GEPIA) of keratocan in pancreatic cancer shows it is significantly upregulated in tumour tissues in comparison to healthy para-cancerous tissues, and patients with pancreatic cancer and high levels of keratocan tend to have low survival rates [166]. Keratocan is also shown to be upregulated in BxPC-3 and PANC-1 pancreatic cell lines exhibiting enhanced proliferation and migration but reduced cellular apoptosis [167]. Furthermore, in pancreatic ductal adenocarcinoma (PDAC), the activity levels of P53 and P21 pathways during cell division were assessed by transfecting PDAC cell lines with pENTER-human keratocan construct, and it was reported that P53 and P21 expression levels were significantly reduced in pancreatic ductal adenocarcinoma cells overexpressing keratocan [168]. These findings suggest that keratocan may play an important role in modulating cell division in pancreatic cancer.

2.1.8 Osteomodulin

Osteomodulin (OMD) also termed osteoadherin (OSAD) is mainly expressed in mineralised tissues such as bones and teeth, and functions in hydroxyapatite binding governing the shape of collagen fibrils, including cell adhesion, tooth formation, and in bone diseases such as osteoarthritis and heterotopic ossification [169]. OMD is upregulated by BMP2, a potent osteoinductive cytokine, that serves roles in the apoptosis and growth of osteoblast cells. Osteomodulin also regulates important cancer pathways such as Ras, P53, Jun/Fos, MYC and NFkB signalling.

Breast cancer metastases often lead to the development and spread of cancer cells to bone, brain, liver and lungs and is the main cause of mortality in women [170]. Osteomodulin (OMD) is associated with breast cancer bone metastases, and immunohistochemical analysis shows heterogeneous localisation of OMD in tumour stromal tissue and infiltrating immune cell [171]. In breast cancer cell lines, OMD does not promote extracellular matrix remodelling, and similarly in bone cancer cell line OMD partially affects osteoblast activity and does not play a role in bone resorption including osteoclast activity in experimental tumour microenvironment [172].

Osteomodulin-expressing cell lines do not affect cell invasion as measured by a Boyden assay with matrigel-coated chambers, but OMD expression increases significantly during cell migration via a scratch wound assay where cell proliferation is prevented by mitomycin pretreatment [172]. Further, in mice model, breast cancer cell lines introduced by intracardiac injection showed increase in the number of cytokeratin-19-positive cells in bone, indicating a role of OMD in colonising the bone cells in vivo [173]. Proteomic analysis reveals OMD-dependent cell migration in vitro and bone colonisation in vivo is dependent on OMD-CDK1 axis that may regulate the OMD-dependent phenotypes, and CDK1 inhibition affects cell viability down streaming regulating OMD in patient-derived breast cancer samples [174].

Bladder cancer also called as urothelial carcinoma is the most common form of urinary tract cancer prevalent among older adults [175]. Osteomodulin (OMD) and proline/arginine-rich end leucine repeat protein (PRELP) are specifically expressed in bladder cancer, and their levels are significantly downregulated in bladder cancers from early stages involving various types of epithelial cells [142]. In vitro studies including gene expression profiling using bladder cancer cell lines revealed OMD application suppresses cancer progression by inhibiting TGF-β and EGF pathways, which has the capacity to reverse epithelial–mesenchymal transition (EMT) and activates cell–cell adhesion while inhibiting various oncogenic pathways [176, 177]. Furthermore, in the bladder epithelia, PRELP knockout gene in a mouse model causes partial EMT, loss of tight junctions of the umbrella-shaped cells and formation of a bladder carcinoma by spontaneous breakdowns of the umbrella cell layer. Data indicates OMD and PRLEP are endogenous inhibitors of cancer initiation and progression as they control EMT pathway and may have potential for the treatment of bladder cancer.

Further, recent findings also suggest that pleomorphic adenoma gene-like 2 (PLAGL-2), a proto-oncogene, functions as a crucial regulator of epithelial–mesenchymal transition (EMT) factor to promote bladder cancer cell invasion and metastasis [144]. Upregulated PLAGL-2 is linked to unfavourable survival, and PLAGL-2 deficiency significantly inhibits the proliferation and metastasis of bladder cancer cells. Furthermore, PLAGL-2 inhibits the Hippo pathway that facilitates the nuclear translocation of YAP1/TAZ, a transcription coregulator involved in cellular proliferation and apoptosis suppression increasing YAP1/TAZ activity mediated by RACGAP1-RhoA activation. Therefore, PLAGL-2/RACGAP1/RhoA GTPase/YAP1 signalling is a potential pathway for bladder cancer therapy.

Gastric cancer, also called as stomach cancer, is the fifth most frequently diagnosed and the third leading cause of cancer deaths worldwide [178]. Gastric carcinoma can be intestinal, well differentiated or diffuse and undifferentiated with distinct morphologic appearances, pathogenesis and genetic profiles [179]. The potential curative treatment approach includes surgical resection with adequate lymphadenectomy, and current evidence suggests perioperative therapies are essential to improve patient survival, thus in advanced unresectable metastatic cases patients can be only offered life-prolonging palliative therapy regimens.

Osteomodulin (OMD) is highly expressed in gastric cancer tissues and can be detected preoperatively to evaluate the prognosis of patients affecting the survival and treatment regime [180]. Preoperative detection of biomarkers that can predict postoperative survival of gastric cancer patients has important implications for surgical procedures, postoperative chemoradiotherapy and follow-up. Thus, multicentre cancer database analysis verified by qRT-PCR and western blotting shows mRNA expression of OMD significantly correlates with age and clinical stages. It also associates with a variety of immune cells, ENTPD1 an immune checkpoint regulator and chemokine CXCL12 tend to affect the occurrences and development of gastric cancer. Therefore, it has been suggested that OMD may become a clinically feasible prognostic biomarker of gastric cancer and a new target for immunotherapy.

2.1.9 Epiphycan

Epiphycan (EPYC) also called as dermatan sulphate proteoglycan-3 (DSPG3), regulates the formation of collagen fibrils and other extracellular matrix proteins [181]. EPYC functions to play a crucial role in a variety of human diseases, such as osteoarthritis, high myopia, rheumatoid arthritis, bone growth and pancreatic cancer, and acts as an independent risk factor in ovarian and prostate cancer. Lack of expression of EPYC can also promote hearing impairment and corneal dystrophy.

Ovarian cancer (OC) is the third most common and malignant tumour of the female reproductive system [182]. Of the epithelial cell-derived ovarian cancer, 90% presents high mortality rates; even after recent advances in drug targets and surgical techniques, the survival rate of ovarian cancer patients is at 30–40% [183]. The risk of developing ovarian cancer in females increases with age typically after 55 years and above, and most of the diagnosis is made at an advanced stage due to lack of specific symptoms at an early age. The main treatment measure includes surgery, chemotherapy, radiotherapy, targeted drugs and hormone therapy [184].

Transcript sequencing of ovarian cancer and associated intestinal metastasis shows epiphycan and lumican are the most upregulated genes, and after complete tumour reduction surgery, only epiphycan expression is found to correlate with the poor prognosis, suggesting a malignant role of epiphycan in the progression of ovarian cancer; further, the expression level of epiphycan in ovarian cancer is higher than in normal ovaries and correlates with the increased level of MMPs, CXCLs and TGF-beta [185]. Gene ontology (Go) analysis shows that EPYC plays a role in the arrangement of extracellular matrix, and the Kyoto Encyclopedia of Genes and Genomes (KEGG) sequencing analysis suggests that epiphycan may be expressed through PI3K/AKT, focal adhesion and integrin signalling pathways. PI3K/AKT pathway is most frequently activated in other types of human cancer and considered as a promising therapeutic target. The axis contributes towards oncogenic transformation involving cell proliferation and survival, metabolic reprogramming, suppressing autophagy, senescence and invasion/metastasis as epithelial mesenchymal transformation [186]. EPYC is suggested to interact with phosphatidylinositol-specific phospholipase-Cγ2 (PLCG2) and cysteine-rich receptor-like kinases (CRK), key signal transduction molecules in cancers [187]. Epiphycan is thus considered to be a potential candidate for diagnosis and treatment of ovarian cancer.

Pancreatic cancer (PC) presents certain clinical features as difficult early diagnosis and rapid progression of tumours, with limited treatment choice including surgical resection. Therefore, to develop instructive biomarkers for early diagnosis and late-stage treatment of pancreatic cancer, thorough integrated bioinformatic analysis indicates epiphycan function as an independent risk factor that can predict poor prognosis in pancreatic cancer [188]. Robust data screening and validation from The Cancer Genome Atlas (TCGA) and gene expression omnibus (GEO) tools using log-rank test and univariate cox regression analysis obtained 39 candidate genes, and subsequently the LASSO regression model-based analysis identified two ultimate genes in pancreatic cancer, namely, the epiphycan and mesenchymal epithelial transition gene (also called as MET gene). Bioinformatic analysis further validated with molecular experiments including RT-qPCR, CCK8, EdU, colony formation assay, western blot, live–dead cell staining and subcutaneous tumour formation assays demonstrates epiphycan is highly expressed in pancreatic cancer cell lines and could promote the proliferation of cells through the PI3K-AKT signalling pathway in vivo and in vitro [189]. Epiphycan is thus suggested to be a neo-biomarker for pancreatic cancer as well as a potent therapeutic target.

Prostate adenocarcinoma (PRAD) is a common malignant tumour in elderly men, and the present screening methods involve the detection of serum prostate-specific antigens, while the treatment includes surgery, radiotherapy, chemotherapy and endocrine therapy [190]. Thus, there is an unmet need to develop a rapid and noninvasive method for predicting the prognosis of diseases in patients with PRAD. Gene expression data derived from The Cancer Genome Atlas (TCGA) programme in patients with prostate adenocarcinoma predicts the level of stromal and immune cell infiltration using the ESTIMATE algorithm that identifies eight individual differentially expressed genes as C6, S100A12, MLC1, PAX5, C7, FAM162B, CAMK1G and TCEAL5 that significantly predicts favourable overall survival, and one differentially expressed gene, namely, the epiphycan, associates with poor overall survival [191]. Data analysis based on patient with high and low score arrays for the median immune cells, stromal scores, Gene Ontology (GO) and Kyoto Encyclopedia of Genes and Genomes (KEGG) pathways show association between differentially expressed genes and the overall survival rate. Furthermore, protein–protein interaction network (PPI) of DEGs constructed using the STRING tool identified 30 genes based on the PPI network as ITGAM, CD4, CD3E, IL-10, LCP2, ITGB2, ZAP-70, C3, CCL19, CXCL13, CXCL9, BTK, CCL21, CD247, CD28, CD3D, FCER1G, PTPRC, TYROBP, CCR5, ITK, CCL13, CCR1, CCR2, CD79B, CYBB, IL2RG, JAK3, PLCG2 and CD19. These prominent gene clusters have the most associations with other matrix regulating genes, indicating crucial roles in the prognosis of PRAD. Epiphycan is the only gene associated with poor survival and therefore can contribute to the prediction and interpretation of prognosis in prostatic adenocarcinoma.

2.1.10 Opticin and Cancer

Opticin (OPTC) was first isolated from the ciliary body of the eyes by non-pigmented ciliary epithelial cells in advanced proliferative retinal disease [192]. OPTC has been detected by mRNA expression analysis additionally in cartilage, brain, ligament, male testis and muscles. OPTC interacts with collagen, growth hormone and heparin sulphate glycosaminoglycans and plays a role in modulating cell anchorage and growth factor sequestration and regulates angiogenesis [193, 194].

Chronic lymphocytic leukaemia (CLL) is the most common adult leukaemia characterised by an accumulation of B-cells expressing cell surface antigens CD5, CD19, CD23, $CD20_{dim}$ and $sIgM_{dim}$ [195]. Opticin is shown to be overexpressed in six cell lines of CLL and mantle cell lymphoma (MCL) patients. Tumour cells, leukaemic cells from blood, bone marrow of haematological malignancies and peripheral blood mononuclear cells (PBMC) from normal healthy donors don't show opticin by RT-qPCR analysis [196]. Further, CLL cells express a smaller molecular size of opticin protein at 37 kDa size, which are not synthesised in normal leukocytes. A similar size 37-kDa OPTC protein with no posttranslational modifications has also been isolated from normal rat model [197]. The 37-kDa CLL opticin detects both the C- and N-terminus sites when treated with anti-OPTC antibody localised in the nucleus and endoplasmic reticulum of the leukaemic cells indicating that the protein may not be a degradation or cleavage product and may represent the core opticin protein without side chain (glycan) modifications. The unique expression of an unglycosylated OPTC with aberrant localisation in CLL cells suggests involvement in the cellular and molecular pathobiology that can be further explored for developing new targeted therapeutics.

Breast cancer angiogenesis involves rapid increase in the formation of blood vessels required for tumour growth, and the breast cancer cells derive nourishment and oxygen supply through the growing network of capillaries present in the system; thus, angiogenesis is a critical factor in the development, progression and metastasis of solid tumour cells [198]. Opticin is a major component of the vitreous humour of the eye which is relatively avascular and acellular structure, and past studies indicate opticin contributes towards inhibition of angiogenesis [199]. Opticin thus suppresses tumour cell proliferation and migration in vitro in breast cancer cell lines MDA-MB-231 and MDA-MB-468 [200], and in in vivo human breast tumour xenografts when applied to mice model intradermally the implanted breast cancer cell grow to 200 mm and after the administration of opticin shows a reduction in vessel growth up to 38% when compared to untreated controls, and these vessels were immunohistochemically evaluated by CD-31 antibody to endothelial cells (blood vessel marker). A microscopic loss in vessel clarity was observed, and similarly adenoviral opticin-treated tumours are significantly smaller than the controls; thus, opticin is suggested to be a potent antiangiogenic and anti-proliferative molecule in vitro and in vivo and could be considered as a novel cancer therapeutic in the treatment of solid tumours [201].

2.1.11 Osteoglycin and Cancer

Osteoglycin (OGN) also called as osteoinductive factor is expressed in various organs and modulates bone formation, collagen fibrillogenesis, tumorigenesis, vascular differentiation and connective tissue-related diseases. OGN in cancerous pathologies is implicated in altering cellular processes linked to tumorigenesis, including cell proliferation, invasion, metastasis and epithelial–mesenchymal transition [202].

Pancreatic cancer associates with limited therapeutic options and poor prognosis. Osteoglycin is upregulated in pancreatic cancer tissues and cell lines as analysed by gene expression profiling interactive analysis (GEPIA), immunohistochemistry, western blotting and quantitative PCR [203]. Moreover, the expression of OGN is closely associated with tumour nodes and metastasis (TNM) stages, where osteoglycin is higher in stage III than in stage I and II pancreatic cancer patients, High levels of OGN correspond with lower survival rates [204] and overexpression of OGN promotes proliferation, migration and invasion of the pancreatic cells inhibiting apoptosis. DNA binding-4 (ID4) protein is upregulated by OGN, and silencing ID4 by small interfering RNA partially reverses the tumour-promoting effect of OGN [202]. Thus, the elevated expression of OGN associates with pancreatic cancer progression and may serve as a potential biomarker for the diagnosis and prognosis by targeting the OGN/ID4 axis.

Breast cancer is a major health problem especially among women, and due to the heterogeneity of the tumour pathogenesis, drug resistance and advanced metastasis in some patients, breast cancer can become untreatable [205]. OGN expression decreases in breast cancer tissue compared to normal control, where decreased osteoglycin level correlates with increase in pathological aggressive tumour grades with poor overall survival chances [206]. In vitro experiments show that OGN overexpressed by plasmid transfection significantly inhibits cell proliferation, colony formation, migration and invasion of breast cancer cell lines. Further, in xenograft tumour models, overexpression of osteoglycin represses the growth of MCF-7 breast cancer cell lines in vivo and alleviates tumour compression on the surrounding structures. OGN expression also reverses epithelial mesenchymal transition (EMT) by repressing the PI3K/Akt/mTOR pathway, which is a major intracellular signalling pathway that responds to the availability of nutrients, hormones and growth factors, which are known factors to play a significant role in breast tumour cell growth and proliferation; thus, the treatment with PI3K/AKT/mTOR pathway activator (740Y-P) reverses the antitumor effects of OGN on breast cancer cells [207].

Colorectal cancer (CRC) is the third most common cause of cancer-related death in males and females worldwide, second to lung cancer. Colon cancer is sporadic, and approximately 5% are due to an inherited genetic mutation with the median age of diagnosis being over 65 years [208].

Osteoglycin in human colorectal cancer is expressed within the epithelial cell membrane and the stromal tissue. Colorectal cancer cell lines HT29 and SW620 overexpressing osteoglycin attenuate cell migration and invasion ability, resulting in a 31% reduction in migration in the HT29 cells decreasing invasions by 37% in the SW620 cell line [209]. Similarly, wound healing assays show OGN overexpression prevented the average invasiveness of HT29 as well as SW620 cell lines. Further, the cell membrane-bound localisation of osteoglycin facilitates the internalisation of epidermal growth factor receptors (EGFR) from the cell membrane into cytoplasm which inhibits the activation of EGFR phosphorylation, impedeing EGFR kinase activation and attenuates the downstream activators Akt and Zeb-1 affecting EGFR dimerisation, internalisation and allows recruitment of Eps15 and epsin1 to EGFR. Reduced Zeb-1 inhibits epithelial mesenchymal transition (EMT) of cancer cells and decreases tumorigenesis [210]. Therefore, overexpression of OGN inhibits cell-proliferative signals by EGFR activity and shows a mechanism linking high levels of OGN with reduced relapse and death in patients with colorectal cancer.

Squamous vaginal cancer is a squamous cell cancer, in the epithelial lining of the vagina that develops slowly, sometimes over many years. Often, they begin as vaginal intraepithelial neoplasia (VAIN), which is a pre-cancerous condition. VAIN is most common in women who have had hysterectomies (removal of the uterus), cervical cancer or cervical pre-cancer [211]. Risk factors include human papillomavirus (HPV), human immunodeficiency virus (HIV), smoking, excessive alcohol consumption and exposure to diethylstilbestrol (DES) drugs. In patients with squamous vaginal cancer (SVC), analysis by two-dimensional electrophoresis (2-DE), SameSpot gel-electrophoresies, matrix-assisted laser desorption/ionisation mass spectrometry (MALDI-TOF-MS) for protein identification and immunohistochemistry revealed cytokeratin-13 and osteoglycin were underexpressed in patient tissue in comparison to normal cervical and vaginal mucosa tissue, while overexpressed mediators were cathepsin-D, γ-catenin, RAB2A, apolipoprotein E, tropomyosin 3 and HSPA8, suggesting the objective role of osteoglycin in vaginal cancers as a biomarker of disease [212].

2.1.12 Tsukushi and Cancer

Tsukushi (*TSKU*) interacts with several vital signalling molecules to serve as a mediator in cell signalling pathways that involves the regulation of bone morphogenetic protein (BMP), fibroblast growth factor (FGF), transforming growth factor-β (TGF-β) and cell viability-related Wnt factor [213]. A few studies have additionally shown the functional significance of TSKU in cancers, and its overexpression increases cell proliferation while inhibiting the epithelial mesenchymal transition (EMT) in lung cancer. TSKU overexpression associates with tumour-infiltrating immune cells in various types of cancer and clinical prognosis. TSKU acts as a potent modulator in the wound healing process via inhibition of TGF-β secretion from macrophages.

Non-small cell lung cancer (NSCLC) is characterised by malignant cancer cells that develop within the lung tissue. About 80% of the lung cancer starts from altered cells grouped together as NSCLC, and their subtypes are as adenocarcinoma, squamous cell carcinoma and large cell carcinoma [214]. Tobacco is the most common cause (approximately 90%) of all lung cancers, and other risk factors include alcohol, environmental exposure to second-hand smoke, asbestos, arsenic, chromium, nickel, exposure to ionising radiation and polycyclic aromatic hydrocarbons [215]. Treatment option can vary based on comorbidities, tumour stage and molecular characteristics of the disease, and treatment includes surgery, chemotherapy, radiation therapy (RT) or a combined modality approach for distant metastases in late stages. In NSCLC, tsukushi is overexpressed and undergoes methylation that regulates tissue-specific gene expression by recruiting proteins involved in either repression or inhibition of transcription factors for DNA; therefore, differentiated cells develop a unique DNA methylation pattern that regulates gene transcription [216]. TSKU expression significantly correlates with low tumour-infiltrating B-cell and poor patient prognosis including survival in NSCLC. TSKU is also highly expressed in lung adenocarcinoma tissues, and in H1975 cell line TSKU shows functional regulation in inhibition of epithelial mesenchymal transition (EMT), including cadherin-1, snail family transcriptional repressor 1, snail family transcriptional repressor 2 and vimentin [217]. Cell counts in xenograft model show TSKU promotes cell proliferation in lung adenocarcinoma. Further, according to the Oncomine and TIMER databases, differential expression of TSKU between tumour and normal tissues is reported for lung, breast, kidney and liver cancer, but only in lung cancer high expression of TSKU significantly associates with poor overall patient survival; therefore, TSKU is suggested to act as a potential independent prognostic biomarker in lung cancer.

Hepatocellular carcinoma (HCC) is the fourth leading cause of cancer-related mortality worldwide, with a 5-year survival rate of less than 14% [218]. Patients with potentially curable HCC disease stage are usually recommended for transplant as treatment option but present limited accuracy in prognosticating post-treatment recurrence and survival [219]. Emphasis on the identification of biomarkers to stratify prognosis and optimise treatment strategies in HCC is prioritised. Thus, a novel circulating tumour cell (CTC)-based transcriptomic profiling assay has been developed that demonstrates translation of a tissue-based messenger RNA into liquid biopsy through noninvasive setting method, and the estimates show HCC-circulating tumour cell mRNA score combined with the Nano-Velcro assay and NanoString nCounter platform can be used for quantifying the circulating tumour cell risk score in HCC [220]. The prognostic HCC-CTC risk score assessed with The Cancer Genome Atlas (TCGA), validated with independent clinical CTC, and HCC-CTC risk score through the integrated data framework identified the top ten prognostic genes in HCC includes tsukushi expression were significant along with discoidin domain receptor tyrosine kinase-1, enoyl-CoA hydratase 3-hydroxyacyl CoA dehydrogenase, androgen receptor, lumican (SLRP), hydroxysteroid 17-beta dehydrogenase-6, prostate transmembrane protein, androgen-induced 1 protein, N-terminal EF-hand calcium binding protein and ladinin-1; while solute carrier family members were highly expressed genes in HCC with low expressions in white blood cells. The top ten identified gene accurately discriminated overall survival in HCC cohort with hazard ratio of 2.0 at 95% confidence interval. The combined use of the scoring system and HCC-CTC risk panel successfully distinguished artificial blood sample spike with an aggressive HCC cell type [221]. Thus, the noninvasive approach allows real-time disease profiling and dynamic prognostication of HCC.

2.1.13 Podocan-Like Protein-1 (PODNL-1) and Cancer

Podocan-like protein-1 isolated from the bone matrix was first identified with advanced computational screening method, and it was found to be highly expressed in glioma, bladder and ovarian cancers representing as a potential prognostic biomarker [222].

In *glioma* high level of PODNL-1 with the presence of low-methylation CpG sites can be coupled with immunotherapeutic targets PD-1, PD-L1 and CTLA-4 [223]. Overexpression of PODNL-1 in glioma cells also promotes mesenchymal-related biomarkers including vimentin, N-cadherin, snail and fibronectin.

Ovarian cancer occurs when abnormal epithelial cells in the ovary begin to proliferate rapidly and that may spread to the fallopian tube or the peritoneum lining the abdominal cavity [224]. The risk of developing ovarian cancer is greatest in women aged between 75 and 79 years with a 5-year survival rate of less than 50% [225]. The main reason behind poor survival is recurrence subsequent to therapy; thus, to improve the prognosis of the diseases, recently investigated novel treatment strategies suggest the role of miR-1908 as a biomarker which controls the gene expression of PODNL-1, along with SIRT-2, SBF-1, JUNB and HOXB-8 in ovarian cancer patient cohort selected from The Cancer Genome Atlas [226]. The analysis of clinical RNA sequence and mRNA sequence shows low expression of these genes predict an improved prognosis, and miR-1908 is able to regulate these candidate genes by binding to the three untranslated regions (3'UTRs) of messenger RNA mediating protein–protein interactions. PODNL-1 is a transcription factor in ovarian cancer, where SET binding factor (SBF-1) encodes a member of the protein-tyrosine phosphatase family that contains a guanine nucleotide exchange factor necessary for cell growth and differentiation [227]. Similarly, sirtuin-2 (SIRT-2) is a nicotinamide adenine dinucleotide (NAD)-dependent deacetylase involved in cell proliferation, death, senescence and stress responses in ovarian cancer. Another transcription factor Jun-B (JUN-B) promotes ovarian tumour metastasis and progression, and silencing JUN-B reduced cell growth and colony formation that associates with decreased protein-1 activity [228]. Low expression of miR-1908 may thus act as a valuable biomarker to identify ovarian cancer patients with a poor prognosis. Potential target genes of the miR-1908 are therefore likely to have functions in tumorigenesis and progression in ovarian cancer.

Gliomas are a diverse group of primary tumours originating from glial cells in the brain and spinal cord within the central nervous system. They range from low-grade pilocytic astrocytomas to malignant glioblastomas [229]. Despite advances in molecular and genetic biomarker research, the diagnostic accuracy lacks due to diffuse and infiltrative pathophysiology of high-grade gliomas and pose challenges in treatment and management. The functional significance of PODNL-1 in glioma progression was assessed by analysing pathological grades II–IV classified by the international standardised *WHO Blue Book*. The Cancer Genome Atlas (TCGA) database for determining the overall survival time between high and low PODNL-1 expressions [230] shows in high-grade glioma, PODNL-1 overexpression qualifies as a survival risk to glioma patients. Furthermore, suppression of PODNL-1 inhibits glioma cell proliferation, migration, epithelial mesenchymal transition (EMT) and induce apoptosis. Conversely, the opposite trend was reported in PODNL-1 overexpressing cells. The effects of cell growth and migration caused by PODNL-1 in a knockdown mouse model could be partially rescued by Akt activator and inhibitor indicating that PODNL-1 aggravated glioma malignant behaviour by regulating the Akt/mTOR pathway, a conserved signal transduction network in cancer cells that promotes cell survival, cell growth and cell cycle progression [231].

Bladder cancer affects the urinary bladder, an organ responsible for storing urine; abnormal cells proliferate within the tissue to form tumours that are more common among older adults. The risk factors include smoking, chemical exposure and chronic bladder inflammation [232]. In developing countries such as Africa, Asia, Caribbean and South America, *schistosomiasis* infection is the most common cause of bladder cancer [233]. *Schistosoma haematobium* is a parasitic flatworm that becomes embedded in the bladder wall during the larva stage, leading to irritation, chronic inflammation, squamous metaplasia and dysplasia progressing as squamous cell carcinoma of the urinary bladder, and worldwide approximately 600 million people in developing countries are at risk of infesting bladder cancer [234].

PODNL-1 plays a significant in the progression of bladder cancer, and data analysis from The Cancer Genome Atlas shows PODNL-1 is higher than the adjacent non-caner tissue region [235], where high expression of PODNL-1 is related to pathological stages, grades and histological categories. In bladder cancer cell lines, PODNL-1 is significantly higher than the normal bladder epithelial cells (SV-HUC-1), and PODNL-1 knockdown decrease the proliferation and migration of cancer cells, while overexpression promotes cancer cell spreading [236]. Statistical weighted gene co-expression network analysis (WGCNA) and functional enrichment analysis indicate a positive correlation between PODNL-1 expression and epithelial mesenchymal transformation (EMT) pathway in bladder cancer. Furthermore, the gene set cancer analysis (GSCA) validates high expression of PODNL-1 activates the EMT pathway [237]. Real-time PCR analysis shows PODNL-1 gene expression level negatively correlates with e-cadherin and positively with N-cadherin expression, thus PODNL-1 also regulates EMT process [238]. As high expression of PODNL-1 associates with poor prognosis in patients, PODNL-1 can act as a potential therapeutic target and prognostic biomarker of bladder cancer.

2.2 SLRPs, Inflammation and Degenerative Bone Disease

SLRPs are an essential component of the skeletal tissue, with defined roles in bone formation stages, that includes cell proliferation, matrix deposition, remodelling and mineral deposition [239]. Bone is a modified connective tissue composed of a three-dimensional mineralised, organic and inorganic matrix and cell types named as the osteoblast (neo-bone cells), osteocytes (mature cells), osteoclast (bone removing cells) and osteoprogenitor cells (bone stem cells) [240]. Bone differs from other tissues as it comprises of organic matrix formed by the type I collagen fibres (90%), and non-collagenous proteins include osteocalcin, osteonectin, osteopontin, fibronectin, bone morphogenetic proteins, transforming growth factors and the SLRPs decorin, biglycan, lumican and osteoadherin [241]. The inorganic matrix of the bone predominantly consists of phosphate, calcium ions with significant amounts of bicarbonate, sodium, potassium, citrate, magnesium, carbonate, fluorite, zinc, barium and strontium [242].

SLRPs correlate with degenerative bone diseases such as osteoporosis, osteoarthritis and ectopic bone formation during depletions or overexpression [243, 244]. These actions are mediated by key signalling molecules that regulate activities of transforming growth factor beta, bone morphogenetic protein, integration of Wnt and nuclear factor kappa-B influencing the availability of osteogenic precursors leading to subsequent development, differentiation and modification of the functional load-bearing capacities of the bone.

The localisation of SLRPs in the extracellular matrix (ECM) allows collagens to regulate the kinetics, assembly and organisation of fibrils in bone, tendon, cartilage, skin and cornea [245]. In bone degeneration and diseases, SLRPs are reported to be sequestered in the matrix impacting the collagen fibrillogenesis process, and thus SLRPs can exist as soluble molecules, especially when released by proteolytic digestion at the injured tissue site [246]. SLRPs interact with growth factors such as TGF-beta, bone morphogenetic protein (BMP), Wnt inducible secreted protein-1 (WISP-1), von Willebrand factor, platelet-derived growth factor (PDGF), tumour necrosis factor (TNF) alpha and insulin-like growth factor-I (IGF-I) [247]. These biological interactions modulate growth factor bioavailability as specific concentration gradients, and the pericellular localisation of the SLRPs further allows various types of matrix molecules to interact with different ligands and cell-surface receptors, thereby inducing cell–matrix-specific interactions pertinent to the disease stage [248].

2.2.1 Decorin, Biglycan and Degenerative Bone Diseases

Class I SLRPs decorin and biglycan are reported to be expressed in mature and immature bones during embryonic development and postnatal bone homeostasis, and they show extensive difference in their spatial localisation, suggesting the unique roles these two SLRPs play at specific sites during events such as osteoid formation and subsequent bone mineralisation [249, 250].

Decorin in immature bone is expressed throughout the osteoid matrix and is associated with osteogenic and non-osteogenic layers of the periosteum [251], whereas in mature bone it is present in the pericellular lacunar matrix, the canaliculi of osteocytes and the matrix immediately adjacent to haversian canals, which are a network of tubular canals in the cortical bone allowing the passage of blood vessels and nerves [252].

SLRPs decorin and biglycan due to its relatively small molecular size are involved in the pathophysiology of degenerative bone disease such as osteoporosis and osteoarthritis [253] that leads to the activation of TGF-β, BMPs, receptor tyrosine kinases (ErbB family members), insulin-like growth factor-1 including initiation of cell signalling events through the toll-like receptors (TLRs). Decorin and biglycan bind to all the three isoforms of TGF-β, and bone cell extract analysis reveals TGF-β is mainly bound to decorin core protein increasing the affinity of TGF-β receptors that influence matrix synthesis, bone formation and resorption to maintain tissue homeostasis [254]. Similarly, lack of biglycan results in reduced response to TGF-β by bone marrow stromal cells, and TGF-β members have a positive feedback effect that in turn upregulates the expression of biglycan mRNA and protein [255]. Upon bone tissue stress or injury, sequestered and immunologically inactive decorin is released to the ECM, and proteolytic breakdown by matrix enzymes releases a soluble decorin peptide that interacts with the toll-like receptors (TLR-2 and TLR-4) as endogenous ligand triggering inflammatory response [256]. This process is also relevant in bone tumour biology, as toll-like receptor signalling in bone tumour cells is reported to transform cell functions. A recent analysis also shows that soluble decorin binds to TLR-2 and TLR-4 on macrophages, leading to enhanced production of the proinflammatory protein for programmed cell death-4 (PDCD-4) and oncomir miR-21 [257]. This causes increased translation of PDCD-4 that decreases anti-inflammatory cytokine interleukin (IL-10). Decorin is also reported to attenuate TGF-β1 signalling pathways in bone tumours and prevents PDCD-4 translational repression to generate a proinflammatory microenvironment as an endogenous inhibitor of TGF-β1 by sequestration [258]. Thus, decorin plays a vital role in inflammatory response activities in bone during ageing, degeneration and tumour progression. *Biglycan* in the developing bone is localised around the osteocyte lacunae, and it may act as a transducer of sheer forces to the bone cell surfaces [259]. In mature bone, biglycan is evenly distributed throughout the matrix and acts as a signalling molecule for immune system activation [260]. The core protein molecule of biglycan directly binds to the members of bone morphogenetic protein (BMP) family enhancing BMP-2 activity, whereas the GAG side chain enhances the binding of BMP-4 to osteoblasts, resulting in smad-1/5/8 phosphorylation and increased expression of osteoblastic markers such as runx-2, osterix, osteopontin (OPN/SPP1), bone sialoprotein (BSP) and osteocalcin (BGLAP) [261, 262]. Biglycan also activates the MAPK pathways (ERK and p38 signalling) promoting osteoblast differentiation [263]. This activation is mediated through the GAG chains, as phosphorylation of Erk is not observed when the GAG-free form of biglycan is presented to pre-osteoblasts or macrophages. Furthermore, biglycan triggers the Wnt pathway by binding to both Wnt3 and the Wnt receptor, LRP6/5 [264]. This process is reported during age-related degeneration of bone, where the GAG chain structure and post-translational modifications are affected, altering signalling pathways and cell–matrix interaction allowing differentiation, proliferation and non-maintenance of the regulatory bone function.

2.2.2 Lumican, Fibromodulin and Osteomodulin in Degenerative Bone Diseases

Lumican plays an essential role in developing osteoblast cells, and in early embryonic mouse model, lumican is mainly expressed at the cartilaginous–bone interface, while in the older embryos lumican expression is prominent within the developing bone matrix [265]. Lumican is significantly secreted by differentiating and mature osteoblastic cells and shows potential to act as a biomarker as it distinguishes proliferating pre-osteoblasts from the differentiating osteoblasts [266]. Lumican inhibits differentiation of osteoclast (bone removing cell), and in vitro investigation reveals lumican reduces bone resorption in a dose-dependent manner; thus, lumican is also an osteoclastogenesis marker [267]. The migration and fusion of pre-osteoclasts and the resorptive activity per osteoclast reduces the presence of lumican affecting stages of osteoclastogenesis through the nuclear factor kappa-B ligand (RANKL)-dependent pathways [268]. Lumican inhibits Akt but not MAP kinases such as JNK, p38 and ERK pathways, and co-treatment with an Akt activator reverses the effect of lumican on osteoclast differentiation; therefore, lumican inhibits osteoclastogenesis by suppressing Akt activity and has an osteoprotective role increasing bone formation and decreasing bone resorption, therefore representing a dual regulatory role in bone metabolism making it a promising therapeutic target for osteoporosis.

Fibromodulin (FMOD) with multiple keratan sulphate side chains is a class II SLRP collagen-binding molecule broadly distributed in the connective tissues. Fibromodulin is highly expressed in cartilages and associated with development as well as tissue degeneration in osteoarthritis (OA) progression especially in knee, ankle and temporomandibular joint [269, 270].

In early embryonic stages of the mouse glenohumeral joints where the limb buds are developing from the mesenchymal cells, fibromodulin is not detectable at the protein level. However, in the mid-embryonic stages, intense FMOD staining begins in the scapular and humeral anlage intra- and pericellularly, and in late embryonic stages, strong fibromodulin signal is found in the matrix surrounding the joints and around proliferating chondrocytes in the epiphyses of humerus and scapula. During postnatal maturation until adulthood, FMOD is detected throughout the developing articular surface and the growth plate in cartilages [271, 272]. Thus, it has been proposed that FMOD has a functional role in the early genesis of articular cartilage.

Fibromodulin protein core molecule during osteoarthritis undergoes systematic breakdown into complex smaller fragments by collagenases, a matrix enzyme that destabilise collagen fibril assembly allowing the smaller fragments to be more liable for receptor activation [273]. Another mechanism by which fibromodulin contributes to bone–cartilage degeneration is by sequestering TGF-β/BMP superfamily members in the matrix preventing their binding to the cell receptors. This action is mediated in combination with biglycan as both of these SLRPs have overlapping and possible compensatory functions in joints [269]. BGN and FMOD double-knockout mice exhibit an earlier onset of OA and present abnormal gait characterised by the decreased flexibility of knee and ankle joints in 3-week-old mice, and similar observations are made with temporomandibular joints. FMOD in inflammatory reaction involves the complementary factor C1q and complement inhibitor factor H, as they directly bind to FMOD protein in patients with laryngeal cartilage–bone cancer. FMOD balances the activation of the classical complement pathway in its intact form and also silences the complement cascade by binding to factor H, but when FMOD is degraded or fragmented, as seen in OA, the N-terminal FMOD segment binds to C1q and activates the complement system as a catabolic inflammatory reaction [274].

Osteomodulin (OMD), also termed as *osteoadherin* (OSAD), is a class II keratan sulphate SLRP expressed in mineralised tissues, including bones and teeth [275]. Osteomodulin binds to hydroxyapatite, regulates collagen fibrils and plays a role in the biological processes involving cell adhesion during tooth formation, bone diseases such as osteoarthritis, osteoporosis and heterotopic ossification [169, 276].

Osteomodulin as a potent osteoinductive cytokine expressed during apoptosis, growth of osteoblast cells and are upregulated by BMP-2 signalling pathway in skeletal and dental tissues [277]. OMD binds to BMP-2 via the tenth and eleventh leucine-rich repeats (LRRs) and forms complexes with BMP receptors, thereby activating downstream SMADs to regulate transcriptional responses [278]. Further, OMD-silenced human dental pulp stem cell line (hDPSC) exhibits lower osteoblastic differentiation ability compared to controls [279]. Similarly, in animal model of osteogenesis, the mRNA level of OMD significantly increases in the early stages of development before gradually decreasing in the late stages, which is evaluated by osteoblastic lineage markers such as alkaline phosphates, type I collagen, bone sialoprotein (BSP), osteocalcin and osteopontin [280]. The analysis of the osteogenic gene expression mechanisms shows OMD activates canonical BMP/SMAD pathway and enhances BMP-2-induced SMAD signalling, leading to the formation of bone which can be inhibited by BMP receptor antagonist. BMPs initiate biological events leading to the production of functional bone tissue, and BMP-2 is a potent stimulator of osteoblastogenesis. Forced expression of OMD increases BMP-2 expression activating the canonical BMP effectors SMAD-1 and SMAD-5. Similarly, suppression of OMD downregulates BMP-2 levels preventing intracellular signal transduction in osteogenesis; thus, a possible feedforward cycle is suggested to exist between OMD and BMP-2 [281].

2.2.3 Epiphycan and Osteoglycin in Degenerative Bone Diseases

Epiphycan (EPN) is expressed in epiphyseal growth plate of cartilage–bone junction during embryonic development [282]. EPN in chick and mouse cartilage developmental model suggests that it plays a key role in chondrocyte differentiation and influences cartilage stability. EPN-deficient mice demonstrate it is not essential for embryonic skeletal development, but they play a critical role in proper bone growth length-wise and also for joint maintenances [283]. Thus, EPN influences femur length in male mice, and when there are defects in the growth plate epiphycan is not expressed by the bone. Furthermore, epiphycan maintains cartilage tissue integrity especially with age, and EPN loss along with the absence of biglycan accelerates osteoarthritis development in mice [284]. Therefore, the rapid progression of OA in the double-deficient mouse suggests two possibilities: epiphycan and biglycan have overlapping functions within cartilage, or epiphycan and biglycan have independent coexisting functions, and when these genes are deleted, they speed up the progression of OA. Animal model also demonstrates collagen loss occurs in the articular cartilage of EPN/BGN double-deficient mice, and epiphycan influences the stability of the collagen matrix in biglycan-deficient background, suggesting that biglycan and epiphycan share a function in maintaining joint stability and integrity.

Osteoglycin first isolated from the bovine bone and was called as osteoinductive factor as it was involved with bone formation [285]. The proteolytically cleaved pro-peptide form of osteoglycin and its keratan sulphate side chain is also called as mimecan, due to its relatively small size. The regulatory factors in bovine and human osteoglycin gene expression are p53 DNA binding site, a UV-responsive E-box and multiple IFN-stimulated response elements (ISREs) [202]. In bone disease, osteoglycin is shown to be involved with pathological progression of osteoarthritis and rheumatoid arthritis, and osteoglycin peptide is released in human articular cartilage after degradation with proteolytic matrix metalloproteases [286]. Moreover, the synovial fluid of patients with osteoarthritis shows upregulated expression of osteoglycin [287]. In rheumatoid arthritis patients, low-level laser therapy as treatment procedure increases osteoglycin level in synovial membrane and cultured osteoblast bone cells [288].

2.2.4 Chondroadherin in Degenerative Bone Disease

Chondroadherin is synthesised by the chondrocyte cells to promote attachment to the cell with the surrounding extracellular matrix. It was first purified and isolated from the bovine cartilage and by the bone osteoblast cells as a minor component of the organic matrix. Chondroadherin also promotes attachment of osteoblastic bone cells to solid-state substrate in culture condition, allowing it to bind with collagen at a significantly higher rate than osteonectin or decorin; thus, chondroadherin possibly plays a role in maintaining bone cell attachment on the collagenous matrix by the integrin alpha-5 beta-1 receptors [289].

Immunocytochemical investigations of the rat femoral head at different ages show a general distribution of chondroadherin at the cartilaginous regions, with greatest immunoreactivity in the growth plates of the bone–cartilage junction [290]. With increasing age, the process of mineralisation expands in bone, and the presence of chondroadherin decreases which is mainly expressed by the cartilage chondrocyte cells. During osteoarthritis of bone, a cleavage fragment of about 25 kDa in rats and 30 kDa in growth plate of human cartilage has been reported, and this finding is consistent with the disappearance of chondroadherin when chondrocyte cells undergo hypertrophy [291], suggesting loss of fragmented chondroadherin peptides. These fragments are recognised by the antibodies reacting with the C-terminal part of the protein core by immunohistochemical and western blot analysis, which is the major cleavage site in degenerative tissues.

2.3 SLRPs, Inflammation and Cartilage Tissue Disorder

Cartilage is located between the spine, bones and the joints allowing flexibility and maintenances of the load-bearing function by the human body [292]. Intervertebral discs (IVDs) are white fibrous cartilages between the adjacent vertebrates [293], while the articular hyaline cartilages are positioned between the joints surfaces [294]. The IVDs owing to its mid-spinal orientation and large size facilitate 70–80% of the load transmission along with the vertebral bodies [295], whereas the articular cartilage compensates with the remaining 20–30% [296]. Cartilages function to permit movements such as flexion, extension, medial and lateral rotation at the joints, and during cartilage tissue disorder, the bones and joints become subjected to unbalanced loading, resulting in abnormal biomechanics through the body. It is well known that biomechanics are crucial in cartilage development, growth and physiochemical regulation of extracellular matrix molecules and maintenance of chondrocyte cells. Thus, inadequate or excessive loading has detrimental effect on cartilage matrix often leading to trauma that induces osteoarthritis, worsened with weak genes and unavoidable age-related changes, Fig. 2.1. Thus, osteoarthritis (OA) in cartilage is defined as deformation of cartilage tissue causing microstructural

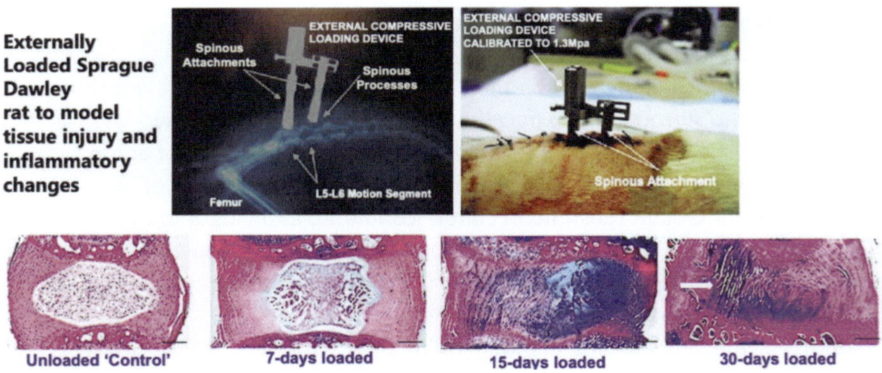

Fig. 2.1 Radiograph shows surgical intervention of the lower lumbar vertebrae (L5–L6 level) on the Sprague Dawley rat with an external compression device to model tissue injury and to understand the inflammatory changes associated with intervertebral disc degeneration. The unloaded control shows no changes in tissue structure, while the 7-, 15- and 30-day loaded rat discs show tissue disruptions which distort the central region of the discs as visualised in 7-day loaded rat disc. These changes facilitate loss of small and large proteoglycan molecules, and in 15-day loaded rat disc an increase in proteoglycan content is visualised as distinct blue staining which indicates an attempted repair process initiated by the disc cells to restore back the proteoglycan content. A 30-day loaded rat disc develops wide fissures (white arrow) due to progressive degenerative changes which associate with inflammatory imbalance. Scale bar 200 μm. (Figure adapted from Dr Polly Lama, Molecular Matrix Lab)

damage and changes in the matrix molecules leading to pain and loss of function [297].

2.3.1 Decorin, Biglycan and Asporin in Cartilage Tissue Disorder

In cartilages, abnormal loading shifts the chondrocytes from normal phenotype to hypertrophy leading to the formations of large cell clusters and overproduction of proinflammatory cytokines IL-6 and IL-1beta; matrix-degrading enzymes MMP-1, MMP-3, MMP-5, MMP-8, MMP-13, ADAMTS-4 and MMP-5; vascular endothelial growth factor; and angiogenin [298]. These catabolic mediators destabilise matrix homeostasis, and overloaded cartilage-joint surfaces show loss of aggrecan, a major proteoglycan molecule that coexists with collagen fibres forming diverse complexes with SLRPs [299], Fig. 2.1. These matrix molecules become proteolytically cleaved into smaller fragments in articular cartilage, and intervertebral discs through combined action of increased mechanical loading and matrix-degrading enzymes which represent a common end stage in cartilage tissue degeneration [300]; further, these fragments interact with cell surface toll-like receptors and integrins to modify cell signalling events.

Decorin is pericellular in localisation in cartilages and are involved in mechanotransduction events regulating matrix micromechanics [301]. Decorin-null mouse shows a significant reduction in pericellular micro-modulus, an intrinsic feature of cartilage stiffness between the matrix molecules [302]. Thus, defects in the pericellular matrix of decorin-null mouse decreases the intracellular calcium level which alters the tissue fluid dynamics and impairs chondrocyte mechanotransduction [303] primarily as the negative fluid charges carried by the aggrecan present in the extracellular matrix are bound to decorin in the articular cartilage [304]. In the human intervertebral discs, several different forms of decorin are present within the annulus fibrosus, nucleus pulposus and cartilage endplate region [305]. In the young intervertebral discs, glycosaminoglycan side chain enriched or glycanated form of decorin is present at a greater proportion of the total proteoglycans in the tissue extracts from annulus fibrosus and cartilage endplate compared to nucleus pulposus region where they are barely detectable. In older discs, these glycanated forms of decorin form only a small proportion of the total proteoglycan content. Further, dermatan sulphate is the predominant glycosaminoglycan side chain attached to decorin in annulus fibrosus [306], whereas chondroitin sulphate side chain is present in cartilage endplate region [307]. Immunohistochemical analysis with antibodies to protein epitopes identifies two non-glycanated forms of decorin in the three regions of the disc, while the non-glycanated form increases with age [308]. N-amino terminal sequence analysis indicates that the non-glycanated form of decorin is possibly the degradation product of the glycanated precursor, as decorin undergoes several post-translational modifications and is cleaved by the matrix degrading enzymes synthesised from the cells of the cartilage tissue [309].

Biglycan triggers catabolic responses in cartilage chondrocyte cells during osteoarthritis and cartilage tissue injury, where biglycan becomes proteolytically cleaved and released to the extracellular matrix to act as a proinflammatory endogenous danger signal molecule that can activate toll-like receptors [310, 311]. Soluble form of biglycan fragments is found in the synovial fluid of patients with cartilage diseases, and biglycan is also expressed and secreted by circulating macrophages releasing inflammasome possibly due to the activation of toll-like receptors 2, 4 and the purinergic P2X receptors through the ERK and NF-κb inflammatory pathways [312]. During the process of inflammation in articular cartilage and synovial fluid, nitric oxide is produced by the chondrocyte cells that accelerate the immune response leading to the recruitment of leukocytes at the area of tissue injury [313] where intact and soluble form of biglycan is present in patients with meniscus tear, early, advanced osteoarthritis (OA) and rheumatoid arthritis (RA) and the concentration of soluble biglycan increases during the late stage of the disease. In intervertebral discs, fragments of biglycan are associated with remodelling of the annulus fibrosus in the vicinity of tissue fissure or tear region and are absent in the gel-like nucleus pulpous region of the discs in age-matched control human tissue and experimental ovine (sheep) animal model [314]. Thus, it is suggested that chondrocytes as well as other immune cells in cartilage diseases generate various types of mediators that alter the biglycan protein core structure releasing intact, soluble and fragmented form that is subsequently released into the synovial fluid of patients with OA as a result of the disease activity. Biglycan is proposed to be a possible diagnostic biomarker in cartilage tissue diseases. The increasing level of intact and soluble biglycan peptides corresponds with the upregulation of its mRNA protein in human osteoarthritis and ovine (sheep) cartilage meniscus investigating osteoarthritis model. Further, synovial fluid of OA and RA patients contains immunoglobulin G autoantibodies against biglycan and other cartilage matrix molecules which suggest their release from matrix along with subsequent loss of immunological tolerance in diseases [315]. Thus, identification of intact, soluble or fragmented form of biglycan in synovial fluid and cartilage tissue diseases along with the activation of endogenous damage-associated molecular pattern-type ligand as the TLR-2, TLR-4 and P2X7/4 demonstrates a mechanism that possibly involves autoinflammation, costimulation and autoimmunity.

Asporin is involved in pathogenesis of osteoarthritis showing increased expression in diseased or injured cartilage tissue when compared to normal tissue [316]. Asporin influences various types of cellular functions in cartilage by binding with TGF-beta receptors, and the ratio of asporin mRNA to TGF-beta mRNA in cartilage and subchondral bone is higher in osteoarthritis (OA) patient group in comparisions to non-OA group positively correlating with the severity of disease [317]. In ATDC5 cell line of in vitro mouse during chondrogenesis, TGF-beta induces phosphorylation of Smad in a dose-dependent manner that counteracts interleukin-1 (IL-1) synthesis suppressing proteoglycan production while stimulating nitric oxide release increasing the expression level of MMPs by upregulating A-disintegrin and metalloproteinase with thrombospondin motifs (ADAMTS) and TIMPs [318]. TGF-beta also activates Runx-2 pathways which lead to hypertrophic differentiation of articular chondrocytes that stimulate the production of ADAMTS and MMP-13 [319]. ADAMTS and MMP-13 degrade the largest proteoglycan aggrecan to which SLRPs bind and effectively cleave type II collagen [320]. Recombinant human asporin inhibits TGF-beta binding to TGF-b RI and RII receptors [48], thus asporin is a potential trigger of OA in cartilage and subchondral bone including injury or ailments accelerating OA involving degradation or breakdown of large structural molecules such as collagen and aggrecan. Past studies also indicate an association between polymorphism in the aspartic acid (D) repeat of the gene encoding asporin (ASPN) and osteoarthritis, but the results are inconsistent among different ethnic groups, and stratified meta-analyses fail to demonstrate a causal relationship between asporin and OA; the reasons for the contradictory findings are yet unknown, but it has been suggested that variation in D allele frequencies may be an important factor as it influence the function of asporin [321].

2.3.2 Lumican, Fibromodulin and Osteomodulin in Cartilage Tissue Disorder

Lumican (LUM), a class II SLRP member, interacts with fibrillar collagens and are involved in host–pathogen interactions where lumican binds with pattern recognition toll-like receptor 4 (TLR-4) that functions as lipopolysaccharide (LPS) sensor [322], resulting in the production of several proinflammatory, antiviral, antibacterial cytokines including fas ligand-mediated apoptosis. Lumican becomes significantly fragmented in osteoarthritis cartilage, but their expression level is reduced to less than 50% of total protein content, however lumican fragments tend to be upregulated in plasma serum and synovial fluid of OA patients, suggesting dysregulation in cartilage, along with the expression of TLR-4 receptors by the chondrocyte indicating inflammatory catabolism in OA [323]. A 'two-hit' model of OA suggests lipopolysaccharide primes the innate immune response that leads to the progression of a full-blown inflammatory cascade facilitating structural changes at the cartilage–joint interface from coexisting complementary factor such as inflammasome activation by damage-associated molecular patterns molecules due to extensive proteolysis. Lipopolysaccharide is considered a hidden risk factor that may explain the association between obesity, abnormal loading, metabolic syndrome and osteoarthritis [324]. Similarly, in the nucleus pulpous region of degenerated intervertebral disc, several SLRP fragments are identified with lumican as an exception, and they tend to be mainly expressed in control tissue; thus, lumican protein could be lost during disc degeneration as the nucleus pulposus region exhibits changes in fibril diameter consistent with tissue fibrosis increasing tissue porosity thus lumican may possibly leach out of the disc tissue [325].

Fibromodulin (FMOD) in cartilage tissue disorder prevents cell adhesion and promotes tissue damage involving inflammatory changes in articular cartilage, where fibromodulin specifically prevents the adhesion of polymorphonuclear neutrophil and fibroblast cells on the articular surfaces of cartilage, and inhibition of cell attachment may be due to the capability of FMOD to mask epitopes of collagen molecules that face the joint cavity [326]. Fibromodulin directly alters the complement component 1q (C1q) and complement inhibitor factor H of the immune system by binding to the protein core molecule of fibromodulin [327]. Complement factor is a part of the innate immune system that regulates cell differentiation, adhesion, migration, activation and survival. Thus, C1q deficiency is associated with autoimmunity as well as increased susceptibility to infections, and complement activation is also elicited when C1q binds to other ligands such as cytokines and DAMPs [328]. N-terminal fragments of fibromodulin in cartilage stimulate interleukin-1 (IL-1) to bind to C1q, but not to factor H [274]. Fibromodulin is suggested to balance the activation of the complement pathway when maintained in its intact form as it silences the complement cascade by binding to factor H. On the contrary, when fibromodulin is degraded or fragmented in osteoarthritis [329, 330], the N-terminal segment binds to C1q eliminatings damaged cells, thus allowing tissue recovery and cell adhesion, and this may lead to uncontrolled inflammation during the late stages of OA.

In intervertebral disc, as the nucleus pulposus region degenerates, fibromodulin expression change due to transcriptional methylation of its core protein structure. During in vitro stress stimulation, low fibromodulin methylation level negatively correlates with its expression. Further, high-throughput sequencing and bioinformatic analysis also show fibromodulin regulates cytoskeletal dynamics and activates integrin pathway by interacting with type I collagen through the leucine-rich repeat domain [331]. As degeneration of nucleus pulposus progress, tissue fibrosis and type I collagen content increases influencing the cytoskeleton dynamics. The nucleus pulposus cells sense changes in the matrix stiffness via the integrin receptor, and the cytoskeleton contracts to alter the cell shape. Fibromodulin promotes stress fibre formation augmenting cell migration and proliferation. Moreover, nucleus pulposus cells in fibromodulin silenced cells inhibit cell proliferation, facilitate apoptosis and increase autophagic bodies; thus, exogenously administered fibromodulin reduces autophagy flux, but past studies also indicate that autophagy activation under hypoxic conditions protects the discs from pressure-induced degeneration [332].

Osteomodulin (OMD), a class II SLRP, is involved in extracellular matrix mineralisation, and OMD downregulation is associated with the sclerotic phenotype leading to abnormal subchondral bone remodelling, sclerosis and calcification of cartilage bone junction, which are the main features of osteoarthritis [333]. Osteomodulin knockout as well as overexpressing male mice and mutant zebrafish model indicates irregularities in its expression play a role in bone and cartilage metabolism leading to the development of osteoarthritis [334, 335]. Underexpression of osteomodulin tend to affect the microarchitecture of bone and cartilage resulting in loss of bone volume, onset of subchondral bone sclerosis and calcification of cartilage with the formation of spontaneous cartilage lesions, where osteomodulin binds to RANKL receptors to inhibit osteoclastogenesis controlling bone remodelling, and these changes impact cartilage homeostasis by altering the regulation of osteoclastogenesis [169, 336]. Thus, targeting OMD could possibly be an approach for controlling osteoarthritis.

In intervertebral disc degeneration, increased expression of osterix, RUNX2, bone morphogenetic protein (BMP) 2, osteomodulin and osteocalcin indicates their role in osteogenic differentiation leading to subsequent calcification of cartilaginous endplates, a poorly understood degenerative change that poses as a significant challenge in the clinical management of patients [337, 338]. Annulus fibrosus and endplate are susceptible to ectopic calcification and share similarities with mineralisation mechanisms as evidenced in other collagen-rich tissues such as articular cartilage, ligaments and skin [339]. Recent investigation also reveals deficiency of progranulin (PGRN), a pleiotropic growth factor, increases tartrate-resistant acid phosphatase (TRAP), cathepsin, collagen X, matrix metallopeptidase 13 (MMP-13) and disintegrin and metalloproteinase with thrombospondin motifs (ADAMTS-5) levels leading to dystrophic calcification, disc narrowing and matrix degeneration breaking down the SLRP molecules.

2.3.3 Epiphycan, Opticin and Osteoglycin in Cartilage Tissue Disorder

Epiphycan (EPN), a chondroitin/dermatan sulphate class III SLRP, is preferentially expressed in epiphyseal cartilage during embryonic development and plays a role in skeletal formation and/or maintenance of tissue structure [340]. Epiphycan in cartilage maintains collagen fibrillogenesis and allows matrix assembly involving cell–matrix interactions by initiating cell signalling events [181]. Epiphycan functions along with biglycan, class II SLRP that signal through toll-like receptors 2 and 4 (TLR-2 and TLR-4) required for repairing localised joint damage where the expression of epiphycan by chondrocytes during tissue injury repair matrix defects but accumulates other metabolic by products throughout life, Fig. 2.1. Thus, Epiphycan loss leads to cartilage degeneration and alters subchondral bone structure [283].

Opticin (OPTC) is expressed in articular cartilage and degraded during osteoarthritis. In vivo, lack of opticin alters the natural balance of other SLRP members which is translated into a protective effect in cartilage leading to overexpression of lumican and biglycan at birth and in adults endows protective effect on the collagen fibres as surface deposition of opticin enables the collagen fibrils to gain resistance against the action of catabolic/proteolytic factors [341]. The protective effect of opticin also decreases the expression of fibromodulin involved in the inflammatory activation of the classical complement system and is associated with progression of cartilage disease; thus, opticin decreases fibromodulin level and favours downregulation of catabolic factors and tissue degradation.

Osteoglycin (OGN), also named mimecan, was first isolated from bovine bone and which was also termed as osteoinductive factor [342]. The co-purified form of osteoglycin binds to bone morphogenetic protein (BMP-2 and BMP-3) giving it the characteristic osteoinductive properties. Osteoglycin and decorin are carrier molecules that are able to hold and release transforming growth factor beta (TGF-β) and bone morphogenetic proteins (BMPs) in the extracellular matrix space and play important roles during skeletal morphogenesis and in late stages of cartilage development. TGF-β enhances cell proliferation in early undifferentiated stages of chondrogenesis, but they inhibit terminal chondrocyte differentiation in cell culture system [343]. Therefore, osteoglycin, BMP-2, BMP-3 and TGF-β2 potentiate a role in cartilage and bone formation. The mRNAs of osteoglycin that bind to transforming growth factor-β and bone morphogenetic proteins are also detected in hypertrophic chondrocytes of developing mouse by in situ hybridisation method [344, 345]. In osteoarthritis of knee, osteoglycin fragment of approximately 20 kDa has been reported which is absent in control non-osteoarthritis samples, and these fragments are not cleaved by MMP-13 proteases, and they do not increase with in vitro MMP-13 digestion indicating an alternate mechanism for the production of osteoglycin fragments in OA [286].

2.3.4 Chondroadherin in Cartilage Tissue Disorder

Chondroadherin (CHAD) is a cartilage matrix protein known to mediate adhesion of chondrocytes to collagen type II via the integrin α2β1 receptor allowing cell spreading [346]. Chondroadherin fragments are present in adolescent idiopathic scoliosis and intervertebral disc degeneration, and which are absent in control non-degenerated tissue [347]. These fragments tend to increase with severity of tissue degeneration, and in vitro analysis of intervertebral disc demonstrates that ADAMTS-4; ADAMTS-5; cathepsins K, B and L; and MMP-3, MMP-7, MMP-12 and MMP-13 are incapable of cleaving chondroadherin and HTRA1 is the only protease capable of generating CHAD fragments. Thus CHAD is suggested to act as a biomarker distinguishing disc degeneration from normal aging.

Northern blot analysis of chondroadherin mRNA shows it is also present in the femoral articular head and costal cartilages. Localisation by immunocytochemistry analysis indicates chondroadherin in developing femoral head increases with age, and are distinctly seen in the territorial matrix of proliferating cells in the growth plate and developing articular cartilage [290]. In situ hybridisation shows strong signals of CHAD in late-stage proliferative cells of the growth plate and articular cartilage, and in the intervertebral disc their expression is restricted to the cells present in the calcified cartilage region adjacent the bone [291]. Thus, high expression of cell binding chondroadherin molecules in cartilage tissue shows the importance of this SLRP in cell growth and proliferation.

CHAD is shown to be present in chondrocyte cells of all ages and associates with the territorial matrix of cartilage tissue. The C-terminal binding sequence of CHAD is localised in the disulphide loops wherean additional cell binding sequence is represented by the C-terminal peptide binding to heparan sulphate sidechains with high affinity [348]. The interaction between CHAD and collagen type II is of functional importance for communication between the cells and the surrounding matrix as this interaction regulates collagen fibril assembly. CHAD also interacts with the N-terminal and C-terminal helical domains of collagen type VI, suggesting that CHAD may have a biomechanical function during the formation of assembled fibrillar collagen network [349]. Thus, CHAD deletion in experimental animal model slows down the development of the load-bearing matrix function, possibly due to alterations in chondrocyte cell signalling process that assembles linkages between the fibrillar network of collagen. CHAD deletion also alters the cross-linking of the collagen fibers and influences the local osmotic swelling and hydraulic permeability of aggrecan with significant reduction in effective indentation modulus (E_{ind}) of the superficial cartilage zone accompanied by increased heterogeneity in diameters of the surface collagen fibrils [350]. Thus, deletion of CHAD affects collagen network in early skeletal development mainly in the superficial region of the cartilage, while in the middle and deep zone changes in the effective indentation modulus are absent, where aggrecan tends to accumulate, thus indicating a complex association between CHAD and aggrecan synthesis.

References

1. Brown JS, Amend SR, Austin RH, Gatenby RA, Hammarlund EU, Pienta KJ. Updating the definition of cancer. Mol Cancer Res. 2023;21(11):1142–7.
2. van Zijl F, Krupitza G, Mikulits W. Initial steps of metastasis: cell invasion and endothelial transmigration. Mutat Res. 2011;728(1–2):23–34.
3. Cao W, Qin K, Li F, Chen W. Comparative study of cancer profiles between 2020 and 2022 using global cancer statistics (GLOBOCAN). J Natl Cancer Cent. 2024;4(2):128–34.
4. Arem H, Loftfield E. Cancer epidemiology: a survey of modifiable risk factors for prevention and survivorship. Am J Lifestyle Med. 2018;12(3):200–10.
5. Henke E, Nandigama R, Ergün S. Extracellular matrix in the tumor microenvironment and its impact on cancer therapy. Front Mol Biosci. 2019;6:160.
6. Anderson NM, Simon MC. The tumor microenvironment. Curr Biol. 2020;30(16):R921–R5.
7. Popova NV, Jücker M. The functional role of extracellular matrix proteins in cancer. Cancers (Basel). 2022;14(1):238.
8. Kessenbrock K, Plaks V, Werb Z. Matrix metalloproteinases: regulators of the tumor microenvironment. Cell. 2010;141(1):52–67.
9. Greten FR, Grivennikov SI. Inflammation and cancer: triggers, mechanisms, and consequences. Immunity. 2019;51(1):27–41.
10. Iozzo RV, Sanderson RD. Proteoglycans in cancer biology, tumour microenvironment and angiogenesis. J Cell Mol Med. 2011;15(5):1013–31.
11. Appunni S, Anand V, Khandelwal M, Gupta N, Rubens M, Sharma A. Small leucine rich proteoglycans (decorin, biglycan and lumican) in cancer. Clin Chim Acta. 2019;491:1–7.
12. Järveläinen H, Sainio A, Wight TN. Pivotal role for decorin in angiogenesis. Matrix Biol. 2015;43:15–26.
13. Nelimarkka L, Salminen H, Kuopio T, Nikkari S, Ekfors T, Laine J, et al. Decorin is produced by capillary endothelial cells in inflammation-associated angiogenesis. Am J Pathol. 2001;158(2):345–53.
14. Salomäki HH, Sainio AO, Söderström M, Pakkanen S, Laine J, Järveläinen HT. Differential expression of decorin by human malignant and benign vascular tumors. J Histochem Cytochem. 2008;56(7):639–46.
15. Järvinen TA, Prince S. Decorin: a growth factor antagonist for tumor growth inhibition. Biomed Res Int. 2015;2015:654765, 1.
16. Patel S, Santra M, McQuillan DJ, Iozzo RV, Thomas AP. Decorin activates the epidermal growth factor receptor and elevates cytosolic Ca2+ in A431 carcinoma cells. J Biol Chem. 1998;273(6):3121–4.
17. Xie C, Mondal DK, Ulas M, Neill T, Iozzo RV. Oncosuppressive roles of decorin through regulation of multiple receptors and diverse signaling pathways. Am J Physiol Cell Physiol. 2022;322(3):C554–C66.

18. Sulochana KN, Fan H, Jois S, Subramanian V, Sun F, Kini RM, et al. Peptides derived from human decorin leucine-rich repeat 5 inhibit angiogenesis. J Biol Chem. 2005;280(30):27935–48.
19. Mohan RR, Tovey JC, Sharma A, Schultz GS, Cowden JW, Tandon A. Targeted decorin gene therapy delivered with adeno-associated virus effectively retards corneal neovascularization in vivo. PLoS One. 2011;6(10):e26432.
20. Bi X, Tong C, Dockendorff A, Bancroft L, Gallagher L, Guzman G, et al. Genetic deficiency of decorin causes intestinal tumor formation through disruption of intestinal cell maturation. Carcinogenesis. 2008;29(7):1435–40.
21. Iozzo RV, Chakrani F, Perrotti D, McQuillan DJ, Skorski T, Calabretta B, et al. Cooperative action of germ-line mutations in decorin and p53 accelerates lymphoma tumorigenesis. Proc Natl Acad Sci USA. 1999;96(6):3092–7.
22. Rizzo GEM, Cabibbo G, Craxì A. Hepatitis B virus-associated hepatocellular carcinoma. Viruses. 2022;14(5):986.
23. Li Y, Gan L, Lu M, Zhang X, Tong X, Qi D, et al. HBx downregulated decorin and decorin-derived peptides inhibit the proliferation and tumorigenicity of hepatocellular carcinoma cells. FASEB J. 2023;37(4):e22871.
24. Buraschi S, Pal N, Tyler-Rubinstein N, Owens RT, Neill T, Iozzo RV. Decorin antagonizes met receptor activity and down-regulates {beta}-catenin and Myc levels. J Biol Chem. 2010;285(53):42075–85.
25. Neill T, Torres A, Buraschi S, Owens RT, Hoek JB, Baffa R, et al. Decorin induces mitophagy in breast carcinoma cells via peroxisome proliferator-activated receptor γ coactivator-1α (PGC-1α) and mitostatin. J Biol Chem. 2014;289(8):4952–68.
26. Zhang L, Liu C, Gao H, Zhou C, Qin W, Wang J, et al. Study on the expression profile and role of decorin in the progression of pancreatic cancer. Aging (Albany NY). 2021;13(11):14989–98.
27. Nastase MV, Young MF, Schaefer L. Biglycan: a multivalent proteoglycan providing structure and signals. J Histochem Cytochem. 2012;60(12):963–75.
28. Kocbek V, Hevir-Kene N, Bersinger NA, Mueller MD, Rižner TL. Increased levels of biglycan in endometriomas and peritoneal fluid samples from ovarian endometriosis patients. Gynecol Endocrinol. 2014;30(7):520–4.
29. Jacobsen F, Kraft J, Schroeder C, Hube-Magg C, Kluth M, Lang DS, et al. Up-regulation of Biglycan is associated with poor prognosis and PTEN deletion in patients with prostate cancer. Neoplasia. 2017;19(9):707–15.
30. Weber CK, Sommer G, Michl P, Fensterer H, Weimer M, Gansauge F, et al. Biglycan is overexpressed in pancreatic cancer and induces G1-arrest in pancreatic cancer cell lines. Gastroenterology. 2001;121(3):657–67.
31. Chen WB, Lenschow W, Tiede K, Fischer JW, Kalthoff H, Ungefroren H. Smad4/DPC4-dependent regulation of biglycan gene expression by transforming growth factor-beta in pancreatic tumor cells. J Biol Chem. 2002;277(39):36118–28.
32. Hu L, Duan YT, Li JF, Su LP, Yan M, Zhu ZG, et al. Biglycan enhances gastric cancer invasion by activating FAK signaling pathway. Oncotarget. 2014;5(7):1885–96.
33. Hu L, Zang MD, Wang HX, Li JF, Su LP, Yan M, et al. Biglycan stimulates VEGF expression in endothelial cells by activating the TLR signaling pathway. Mol Oncol. 2016;10(9):1473–84.
34. Xing X, Gu X, Ma T, Ye H. Biglycan up-regulated vascular endothelial growth factor (VEGF) expression and promoted angiogenesis in colon cancer. Tumour Biol. 2015;36(3):1773–80.
35. Diehl V, Huber LS, Trebicka J, Wygrecka M, Iozzo RV, Schaefer L. The role of decorin and biglycan signaling in tumorigenesis. Front Oncol. 2021;11:801801.
36. Aprile G, Avellini C, Reni M, Mazzer M, Foltran L, Rossi D, et al. Biglycan expression and clinical outcome in patients with pancreatic adenocarcinoma. Tumour Biol. 2013;34(1):131–7.
37. Liu S, Yan X, Guo J, An H, Li X, Yang L, et al. Periodontal ligament-associated protein-1 knockout mice regulate the differentiation of osteoclasts and osteoblasts through TGF-β1/Smad signaling pathway. J Cell Physiol. 2024;239(3):e31062.

38. Wang L, Wu H, Zhang H, Lu J, Liang Z, Liu T. Asporin promotes pancreatic cancer cell invasion and migration by regulating the epithelial-to-mesenchymal transition (EMT) through both autocrine and paracrine mechanisms. Cancer Lett. 2017;398:24–36.
39. Fan R, Yan X, Zhang W. Relationship between asporin and extracellular matrix behavior: a literature review. Medicine (Baltimore). 2022;101(51):e32490.
40. Wu H, Jing X, Cheng X, He Y, Hu L, Ye F, et al. Asporin enhances colorectal cancer metastasis through activating the EGFR/src/cortactin signaling pathway. Oncotarget. 2016;7(45):73402–13.
41. Yashiro M, Hasegawa T, Yamamoto Y, Tsujio G, Nishimura S, Sera T, et al. Asporin expression on stromal cells and/or cancer cells might be a useful prognostic marker in patients with diffuse-type gastric cancer. Eur Surg Res. 2021;62(1):53–60.
42. Li H, Zhang Z, Chen L, Sun X, Zhao Y, Guo Q, et al. Cytoplasmic Asporin promotes cell migration by regulating TGF-β/Smad2/3 pathway and indicates a poor prognosis in colorectal cancer. Cell Death Dis. 2019;10(2):109.
43. Ding Q, Zhang M, Liu C. Asporin participates in gastric cancer cell growth and migration by influencing EGF receptor signaling. Oncol Rep. 2015;33(4):1783–90.
44. Zhan S, Li J, Ge W. Multifaceted roles of asporin in cancer: current understanding. Front Oncol. 2019;9:948.
45. Rochette A, Boufaied N, Scarlata E, Hamel L, Brimo F, Whitaker HC, et al. Asporin is a stromally expressed marker associated with prostate cancer progression. Br J Cancer. 2017;116(6):775–84.
46. Lall SP, Alsafwani ZW, Batra SK, Seshacharyulu P. ASPORIN: a root of the matter in tumors and their host environment. Biochim Biophys Acta Rev Cancer. 2024;1879(1):189029.
47. Simkova D, Kharaishvili G, Korinkova G, Ozdian T, Suchánková-Kleplová T, Soukup T, et al. The dual role of asporin in breast cancer progression. Oncotarget. 2016;7(32):52045–60.
48. Maris P, Blomme A, Palacios AP, Costanza B, Bellahcène A, Bianchi E, et al. Asporin is a fibroblast-derived TGF-β1 inhibitor and a tumor suppressor associated with good prognosis in breast cancer. PLoS Med. 2015;12(9):e1001871.
49. Giatagana EM, Berdiaki A, Tsatsakis A, Tzanakakis GN, Nikitovic D. Lumican in carcinogenesis-revisited. Biomolecules. 2021;11(9):16.
50. Matsuda Y, Yamamoto T, Kudo M, Kawahara K, Kawamoto M, Nakajima Y, et al. Expression and roles of lumican in lung adenocarcinoma and squamous cell carcinoma. Int J Oncol. 2008;33(6):1177–85.
51. Cappellesso R, Millioni R, Arrigoni G, Simonato F, Caroccia B, Iori E, et al. Lumican is overexpressed in lung adenocarcinoma pleural effusions. PLoS One. 2015;10(5):e0126458.
52. Park HJ, Cha YJ, Kim SH, Kim A, Kim EY, Chang YS. Keratinization of lung squamous cell carcinoma is associated with poor clinical outcome. Tuberc Respir Dis (Seoul). 2017;80(2):179–86.
53. Watanabe T, Miura T, Degawa Y, Fujita Y, Inoue M, Kawaguchi M, et al. Comparison of lung cancer cell lines representing four histopathological subtypes with gene expression profiling using quantitative real-time PCR. Cancer Cell Int. 2010;10:2.
54. Guo Z, Li Z, Chen M, Qi X, Sun Z, Wu S, et al. Multi-omics analysis reveals the prognostic and tumor micro-environmental value of lumican in multiple cancer types. Front Mol Biosci. 2023;10:1158747.
55. Brézillon S, Pietraszek K, Maquart FX, Wegrowski Y. Lumican effects in the control of tumour progression and their links with metalloproteinases and integrins. FEBS J. 2013;280(10):2369–81.
56. Yamamoto T, Matsuda Y, Kawahara K, Ishiwata T, Naito Z. Secreted 70kDa lumican stimulates growth and inhibits invasion of human pancreatic cancer. Cancer Lett. 2012;320(1):31–9.
57. Li X, Truty MA, Kang Y, Chopin-Laly X, Zhang R, Roife D, et al. Extracellular lumican inhibits pancreatic cancer cell growth and is associated with prolonged survival after surgery. Clin Cancer Res. 2014;20(24):6529–40.
58. Lu YP, Ishiwata T, Kawahara K, Watanabe M, Naito Z, Moriyama Y, et al. Expression of lumican in human colorectal cancer cells. Pathol Int. 2002;52(8):519–26.

59. Radwanska A, Litwin M, Nowak D, Baczynska D, Wegrowski Y, Maquart FX, et al. Overexpression of lumican affects the migration of human colon cancer cells through up-regulation of gelsolin and filamentous actin reorganization. Exp Cell Res. 2012;318(18):2312–23.
60. Watanabe T, Komuro Y, Kiyomatsu T, Kanazawa T, Kazama Y, Tanaka J, et al. Prediction of sensitivity of rectal cancer cells in response to preoperative radiotherapy by DNA microarray analysis of gene expression profiles. Cancer Res. 2006;66(7):3370–4.
61. Coulson-Thomas VJ, Coulson-Thomas YM, Gesteira TF, Andrade de Paula CA, Carneiro CR, Ortiz V, et al. Lumican expression, localization and antitumor activity in prostate cancer. Exp Cell Res. 2013;319(7):967–81.
62. Nikitovic D, Papoutsidakis A, Karamanos NK, Tzanakakis GN. Lumican affects tumor cell functions, tumor-ECM interactions, angiogenesis and inflammatory response. Matrix Biol. 2014;35:206–14.
63. Vecchiotti D, Clementi L, Cornacchia E, Di Vito NM, Verzella D, Capece D, et al. Evidence of the link between stroma remodeling and prostate cancer prognosis. Cancers (Basel). 2024;16(18):3215.
64. Karamanou K, Franchi M, Piperigkou Z, Perreau C, Maquart FX, Vynios DH, et al. Lumican effectively regulates the estrogen receptors-associated functional properties of breast cancer cells, expression of matrix effectors and epithelial-to-mesenchymal transition. Sci Rep. 2017;7:45138.
65. Miziak P, Baran M, Błaszczak E, Przybyszewska-Podstawka A, Kałafut J, Smok-Kalwat J, et al. Estrogen receptor signaling in breast cancer. Cancers (Basel). 2023;15(19):Article 4689.
66. Mangani S, Piperigkou Z, Koletsis NE, Ioannou P, Karamanos NK. Estrogen receptors and extracellular matrix: the critical interplay in cancer development and progression. FEBS J. 2024;292:1558.
67. Karamanou K, Franchi M, Vynios D, Brézillon S. Epithelial-to-mesenchymal transition and invadopodia markers in breast cancer: Lumican a key regulator. Semin Cancer Biol. 2020;62:125–33.
68. Pietraszek K, Chatron-Colliet A, Brézillon S, Perreau C, Jakubiak-Augustyn A, Krotkiewski H, et al. Lumican: a new inhibitor of matrix metalloproteinase-14 activity. FEBS Lett. 2014;588(23):4319–24.
69. Karamanou K, Franchi M, Proult I, Rivet R, Vynios D, Brézillon S. Lumican inhibits in vivo melanoma metastasis by altering matrix-effectors and Invadopodia markers. Cells. 2021;10(4):841.
70. Bielack S, Carrle D, Casali PG, On behalf of the ESMO Guidelines Working Group. Osteosarcoma: ESMO clinical recommendations for diagnosis, treatment and follow-up. Ann Oncol. 2009;20(Suppl 4):137–9.
71. Nikitovic D, Berdiaki A, Zafiropoulos A, Katonis P, Tsatsakis A, Karamanos NK, et al. Lumican expression is positively correlated with the differentiation and negatively with the growth of human osteosarcoma cells. FEBS J. 2008;275(2):350–61.
72. Nikitovic D, Chalkiadaki G, Berdiaki A, Aggelidakis J, Katonis P, Karamanos NK, et al. Lumican regulates osteosarcoma cell adhesion by modulating TGFβ2 activity. Int J Biochem Cell Biol. 2011;43(6):928–35.
73. Nikitovic D, Katonis P, Tsatsakis A, Karamanos NK, Tzanakakis GN. Lumican, a small leucine-rich proteoglycan. IUBMB Life. 2008;60(12):818–23.
74. Berdiaki A, Giatagana EM, Tzanakakis G, Nikitovic D. The landscape of small leucine-rich proteoglycan impact on cancer pathogenesis with a focus on Biglycan and Lumican. Cancers (Basel). 2023;15(14):3549.
75. Davis LE, Shalin SC, Tackett AJ. Current state of melanoma diagnosis and treatment. Cancer Biol Ther. 2019;20(11):1366–79.
76. Brézillon S, Radwanska A, Zeltz C, Malkowski A, Ploton D, Bobichon H, et al. Lumican core protein inhibits melanoma cell migration via alterations of focal adhesion complexes. Cancer Lett. 2009;283(1):92–100.

77. Pietraszek K, Brézillon S, Perreau C, Malicka-Błaszkiewicz M, Maquart FX, Wegrowski Y. Lumican – derived peptides inhibit melanoma cell growth and migration. PLoS One. 2013;8(10):e76232.
78. Sifaki M, Assouti M, Nikitovic D, Krasagakis K, Karamanos NK, Tzanakakis GN. Lumican, a small leucine-rich proteoglycan substituted with keratan sulfate chains is expressed and secreted by human melanoma cells and not normal melanocytes. IUBMB Life. 2006;58(10):606–10.
79. Vuillermoz B, Khoruzhenko A, D'Onofrio MF, Ramont L, Venteo L, Perreau C, et al. The small leucine-rich proteoglycan lumican inhibits melanoma progression. Exp Cell Res. 2004;296(2):294–306.
80. Radwanska A, Baczynska D, Nowak D, Brézillon S, Popow A, Maquart FX, et al. Lumican affects actin cytoskeletal organization in human melanoma A375 cells. Life Sci. 2008;83(19–20):651–60.
81. Jeanne A, Untereiner V, Perreau C, Proult I, Gobinet C, Boulagnon-Rombi C, et al. Lumican delays melanoma growth in mice and drives tumor molecular assembly as well as response to matrix-targeted TAX2 therapeutic peptide. Sci Rep. 2017;7(1):7700.
82. Zheng Z, Granado HS, Li C. Fibromodulin, a multifunctional matricellular modulator. J Dent Res. 2023;102(2):125–34.
83. Pourhanifeh MH, Mohammadi R, Noruzi S, Hosseini SA, Fanoudi S, Mohamadi Y, et al. The role of fibromodulin in cancer pathogenesis: implications for diagnosis and therapy. Cancer Cell Int. 2019;19:157.
84. Kipps TJ, Stevenson FK, Wu CJ, Croce CM, Packham G, Wierda WG, et al. Chronic lymphocytic leukaemia. Nat Rev Dis Primers. 2017;3:16096.
85. Hassan DA, Samy RM, Abd-Elrahim OT, Salib CS. Study of fibromodulin gene expression in B-cell chronic lymphocytic leukemia. J Egypt Natl Canc Inst. 2011;23(1):11–5.
86. Mikaelsson E, Danesh-Manesh AH, Lüppert A, Jeddi-Tehrani M, Rezvany MR, Sharifian RA, et al. Fibromodulin, an extracellular matrix protein: characterization of its unique gene and protein expression in B-cell chronic lymphocytic leukemia and mantle cell lymphoma. Blood. 2005;105(12):4828–35.
87. Mayr C, Bund D, Schlee M, Moosmann A, Kofler DM, Hallek M, et al. Fibromodulin as a novel tumor-associated antigen (TAA) in chronic lymphocytic leukemia (CLL), which allows expansion of specific CD8+ autologous T lymphocytes. Blood. 2005;105(4):1566–73.
88. Klein U, Tu Y, Stolovitzky GA, Mattioli M, Cattoretti G, Husson H, et al. Gene expression profiling of B cell chronic lymphocytic leukemia reveals a homogeneous phenotype related to memory B cells. J Exp Med. 2001;194(11):1625–38.
89. di Celle PF, Carbone A, Marchis D, Zhou D, Sozzani S, Zupo S, et al. Cytokine gene expression in B-cell chronic lymphocytic leukemia: evidence of constitutive interleukin-8 (IL-8) mRNA expression and secretion of biologically active IL-8 protein. Blood. 1994;84(1):220–8.
90. Wu W, Klockow JL, Zhang M, Lafortune F, Chang E, Jin L, et al. Glioblastoma multiforme (GBM): an overview of current therapies and mechanisms of resistance. Pharmacol Res. 2021;171:105780.
91. Mondal B, Patil V, Shwetha SD, Sravani K, Hegde AS, Arivazhagan A, et al. Integrative functional genomic analysis identifies epigenetically regulated fibromodulin as an essential gene for glioma cell migration. Oncogene. 2017;36(1):71–83.
92. Sengupta S, Mondal M, Prasasvi KR, Mukherjee A, Magod P, Urbach S, et al. Differentiated glioma cell-derived fibromodulin activates integrin-dependent notch signaling in endothelial cells to promote tumor angiogenesis and growth. elife 2022;11.
93. Gu HB, Sun L, Liu YX, Li N. Effect of silencing FMOD on Proliferation and Migration of H322 cells and its mechanism. Sichuan Da Xue Xue Bao Yi Xue Ban. 2020;51(2):165–70.
94. Barsouk A, Padala SA, Vakiti A, Mohammed A, Saginala K, Thandra KC, et al. Epidemiology, staging and management of prostate cancer. Med Sci (Basel). 2020;8(3):28.
95. Silva T, Gomes CP, Voigt DD, de Souza RB, Medeiros K, Cosentino NL, et al. Fibromodulin gene variants (FMOD) as potential biomarkers for prostate cancer and benign prostatic hyperplasia. Dis. Markers. 2022;2022:5215247.

96. Nassar ZD, Hill MM, Parton RG, Francois M, Parat MO. Non-caveolar caveolin-1 expression in prostate cancer cells promotes lymphangiogenesis. Oncoscience. 2015;2(7):635–45.
97. Thompson TC, Tahir SA, Li L, Watanabe M, Naruishi K, Yang G, et al. The role of caveolin-1 in prostate cancer: clinical implications. Prostate Cancer Prostatic Dis. 2010;13(1):6–11.
98. Jaggupilli A, Elkord E. Significance of CD44 and CD24 as cancer stem cell markers: an enduring ambiguity. Clin Dev Immunol. 2012;2012:708036.
99. Zhao R, Bei X, Yang B, Wang X, Jiang C, Shi F, et al. Endothelial cells promote metastasis of prostate cancer by enhancing autophagy. J Exp Clin Cancer Res. 2018;37(1):221.
100. Bettin A, Reyes I, Reyes N. Gene expression profiling of prostate cancer-associated genes identifies fibromodulin as potential novel biomarker for prostate cancer. Int J Biol Markers. 2016;31(2):e153–62.
101. Yang Q, Ciebiera M, Bariani MV, Ali M, Elkafas H, Boyer TG, et al. Comprehensive review of uterine fibroids: developmental origin, pathogenesis, and treatment. Endocr Rev. 2022;43(4):678–719.
102. Norian JM, Owen CM, Taboas J, Korecki C, Tuan R, Malik M, et al. Characterization of tissue biomechanics and mechanical signaling in uterine leiomyoma. Matrix Biol. 2012;31(1):57–65.
103. Levens E, Luo X, Ding L, Williams RS, Chegini N. Fibromodulin is expressed in leiomyoma and myometrium and regulated by gonadotropin-releasing hormone analogue therapy and TGF-beta through Smad and MAPK-mediated signalling. Mol Hum Reprod. 2005;11(7):489–94.
104. Pan Q, Luo X, Chegini N. microRNA 21: response to hormonal therapies and regulatory function in leiomyoma, transformed leiomyoma and leiomyosarcoma cells. Mol Hum Reprod. 2010;16(3):215–27.
105. Rogers R, Norian J, Malik M, Christman G, Abu-Asab M, Chen F, et al. Mechanical homeostasis is altered in uterine leiomyoma. Am J Obstet Gynecol. 2008;198(4):474.e1–11.
106. Malik M, Britten J, Segars J, Catherino WH. Leiomyoma cells in 3-dimensional cultures demonstrate an attenuated response to fasudil, a rho-kinase inhibitor, when compared to 2-dimensional cultures. Reprod Sci. 2014;21(9):1126–38.
107. Rudin CM, Brambilla E, Faivre-Finn C, Sage J. Small-cell lung cancer. Nat Rev Dis Primers. 2021;7(1):3.
108. Montanino A, Manzo A, Carillio G, Palumbo G, Esposito G, Sforza V, et al. Angiogenesis inhibitors in small cell lung cancer. Front Oncol. 2021;11:655316.
109. Ao Z, Yu S, Qian P, Gao W, Guo R, Dong X, et al. Tumor angiogenesis of SCLC inhibited by decreased expression of FMOD via downregulating angiogenic factors of endothelial cells. Biomed Pharmacother. 2017;87:539–47.
110. Gao C, Shen Y, Jin F, Miao Y, Qiu X. Cancer stem cells in small cell lung cancer cell line H446: higher dependency on oxidative phosphorylation and mitochondrial substrate-level phosphorylation than non-stem cancer cells. PLoS One. 2016;11(5):e0154576.
111. Jian J, Zheng Z, Zhang K, Rackohn TM, Hsu C, Levin A, et al. Fibromodulin promoted in vitro and in vivo angiogenesis. Biochem Biophys Res Commun. 2013;436(3):530–5.
112. Ma S. Experimental models for preclinical cancer research. Exp Cell Res. 2023;429(1):113643.
113. Jarrett AM, Lima EABF, Hormuth DA, McKenna MT, Feng X, Ekrut DA, et al. Mathematical models of tumor cell proliferation: a review of the literature. Expert Rev Anticancer Ther. 2018;18(12):1271–86.
114. Oldberg A, Kalamajski S, Salnikov AV, Stuhr L, Mörgelin M, Reed RK, et al. Collagen-binding proteoglycan fibromodulin can determine stroma matrix structure and fluid balance in experimental carcinoma. Proc Natl Acad Sci USA. 2007;104(35):13966–71.
115. Rydell-Törmänen K, Andréasson K, Hesselstrand R, Westergren-Thorsson G. Absence of fibromodulin affects matrix composition, collagen deposition and cell turnover in healthy and fibrotic lung parenchyma. Sci Rep. 2014;4:6383.
116. Łukasiewicz S, Czeczelewski M, Forma A, Bąj J, Sitarz R, Stanisławek A. Breast cancer-epidemiology, risk factors, classification, prognostic markers, and current treatment strategies-An updated review. Cancers (Basel). 2021;13(17):4287.

117. Feng Y, Spezia M, Huang S, Yuan C, Zeng Z, Zhang L, et al. Breast cancer development and progression: risk factors, cancer stem cells, signaling pathways, genomics, and molecular pathogenesis. Genes Dis. 2018;5(2):77–106.
118. Ji H, Hu C, Yang X, Liu Y, Ji G, Ge S, et al. Lymph node metastasis in cancer progression: molecular mechanisms, clinical significance and therapeutic interventions. Signal Transduct Target Ther. 2023;8(1):367.
119. Nikitovic D, Kouvidi K, Voudouri K, Berdiaki A, Karousou E, Passi A, et al. The motile breast cancer phenotype roles of proteoglycans/glycosaminoglycans. Biomed Res Int. 2014;2014:124321.
120. Dawoody Nejad L, Biglari A, Annese T, Ribatti D. Recombinant fibromodulin and decorin effects on NF-κB and TGFβ1 in the 4T1 breast cancer cell line. Oncol Lett. 2017;13(6):4475–80.
121. Zheng Z, Nguyen C, Zhang X, Khorasani H, Wang JZ, Zara JN, et al. Delayed wound closure in fibromodulin-deficient mice is associated with increased TGF-β3 signaling. J Invest Dermatol. 2011;131(3):769–78.
122. Lee YH, Schiemann WP. Fibromodulin suppresses nuclear factor-kappaB activity by inducing the delayed degradation of IKBA via a JNK-dependent pathway coupled to fibroblast apoptosis. J Biol Chem. 2011;286(8):6414–22.
123. Khan FU, Owusu-Tieku NYG, Dai X, Liu K, Wu Y, Tsai HI, et al. Wnt/β-catenin pathway-regulated Fibromodulin expression is crucial for breast cancer metastasis and inhibited by aspirin. Front Pharmacol. 2019;10:1308.
124. Xu E, Hu M, Liu Y. Aspirin inhibits proliferation and metastasis of canine mammary gland tumor cells through Wnt signaling axis. Transl Cancer Res. 2021;10(2):589–601.
125. Siegel RL, Wagle NS, Cercek A, Smith RA, Jemal A. Colorectal cancer statistics, 2023. CA Cancer J Clin. 2023;73(3):233–54.
126. Alsheridah N, Akhtar S. Diet, obesity and colorectal carcinoma risk: results from a national cancer registry-based middle-eastern study. BMC Cancer. 2018;18(1):1227.
127. Brenner H, Jansen L, Ulrich A, Chang-Claude J, Hoffmeister M. Survival of patients with symptom- and screening-detected colorectal cancer. Oncotarget. 2016;7(28):44695–704.
128. Berg KCG, Eide PW, Eilertsen IA, Johannessen B, Bruun J, Danielsen SA, et al. Multi-omics of 34 colorectal cancer cell lines – a resource for biomedical studies. Mol Cancer. 2017;16(1):116.
129. Karlsson S, Nyström H. The extracellular matrix in colorectal cancer and its metastatic settling – alterations and biological implications. Crit Rev Oncol Hematol. 2022;175:103712.
130. Deng T, Hou Y, Lin G, Feng C, Liu K, Chen W, et al. A novel Fibromodulin antagonist peptide RP4 exerts antitumor effects on colorectal cancer. Pharmaceutics. 2023;15(3):944.
131. Lin K, Chowdhury S, Zeineddine MA, Zeineddine FA, Hornstein NJ, Villarreal OE, et al. Identification of colorectal cancer cell Stemness from single-cell RNA sequencing. Mol Cancer Res. 2024;22(4):337–46.
132. Oria VO, Zhang H, Zito CR, Rane CK, Ma XY, Provance OK, et al. Coupled fibromodulin and SOX2 signaling as a critical regulator of metastatic outgrowth in melanoma. Cell Mol Life Sci. 2022;79(7):377.
133. Bengtsson E, Neame PJ, Heinegård D, Sommarin Y. The primary structure of a basic leucine-rich repeat protein, PRELP, found in connective tissues. J Biol Chem. 1995;270(43):25639–44.
134. Mikaelsson E, Österborg A, Jeddi-Tehrani M, Kokhaei P, Ostadkarampour M, Hadavi R, et al. A proline/arginine-rich end leucine-rich repeat protein (PRELP) variant is uniquely expressed in chronic lymphocytic leukemia cells. PLoS One. 2013;8(6):e67601.
135. Hong R, Gu J, Niu G, Hu Z, Zhang X, Song T, et al. PRELP has prognostic value and regulates cell proliferation and migration in hepatocellular carcinoma. J Cancer. 2020;11(21):6376–89.
136. Patresan J, Patel H, Chandrasekaran K, Reynolds G. Current treatment paradigm and approach to advanced hepatocellular carcinoma. Cureus. 2024;16(12):e75471.
137. Roy AM, Iyer R, Chakraborty S. The extracellular matrix in hepatocellular carcinoma: mechanisms and therapeutic vulnerability. Cell Rep Med. 2023;4(9):101170.

138. Kosuge H, Nakakido M, Nagatoishi S, Fukuda T, Bando Y, Ohnuma SI, et al. Proteomic identification and validation of novel interactions of the putative tumor suppressor PRELP with membrane proteins including IGFI-R and p75NTR. J Biol Chem. 2021;296:100278.
139. Bellmunt J, Mottet N, De Santis M. Urothelial carcinoma management in elderly or unfit patients. EJC Suppl. 2016;14(1):1–20.
140. Akbas P, Bektas S, Yazici G. The association between variant histology and prognostic, histomorphological and clinical aspects of bladder urothelial carcinoma. Ann Diagn Pathol. 2024;73:152373.
141. Hu X, Li G, Wu S. Advances in diagnosis and therapy for bladder cancer. Cancers (Basel). 2022;14(13):3181.
142. Papadaki V, Asada K, Watson JK, Tamura T, Leung A, Hopkins J, et al. Two secreted proteoglycans, activators of urothelial cell-cell adhesion, negatively contribute to bladder cancer initiation and progression. Cancers (Basel). 2020;12(11):3362.
143. Shozu K, Kaneko S, Shinkai N, Dozen A, Kosuge H, Nakakido M, et al. Repression of the PRELP gene is relieved by histone deacetylase inhibitors through acetylation of histone H2B lysine 5 in bladder cancer. Clin Epigenetics. 2022;14(1):147.
144. Chen H, Yang W, Li Y, Ji Z. PLAGL2 promotes bladder cancer progression via RACGAP1/RhoA GTPase/YAP1 signaling. Cell Death Dis. 2023;14(7):433.
145. Byroju VV, Nadukkandy AS, Cordani M, Kumar LD. Retinoblastoma: present scenario and future challenges. Cell Commun Signal. 2023;21(1):226.
146. Akdeniz Odemis D, Kebudi R, Bayramova J, Kilic Erciyas S, Kuru Turkcan G, Tuncer SB, et al. RB1 gene mutations and genetic spectrum in retinoblastoma cases. Medicine (Baltimore). 2023;102(36):e35068.
147. Dimaras H, Corson TW, Cobrinik D, White A, Zhao J, Munier FL, et al. Retinoblastoma. Nat Rev Dis Primers. 2015;1:15021.
148. Abramson DH. Retinoblastoma: diagnosis and management. CA Cancer J Clin. 1982;32(3):130–40.
149. Hopkins J, Asada K, Leung A, Papadaki V, Davaapil H, Morrison M, et al. Regulates cell-cell adhesion and EMT and inhibits retinoblastoma progression. Cancers (Basel). 2022;14(19):4926.
150. Collin GB, Gogna N, Chang B, Damkham N, Pinkney J, Hyde LF, et al. Mouse models of inherited retinal degeneration with photoreceptor cell loss. Cells. 2020;9(4):4926.
151. Davaapil H, Hopkins J, Bonnin N, Papadaki V, Leung A, Kosuge H, et al. PRELP secreted from mural cells protects the function of blood brain barrier through regulation of endothelial cell-cell integrity. Front Cell Dev Biol. 2023;11:1147625.
152. Adamska A, Domenichini A, Falasca M. Pancreatic ductal adenocarcinoma: current and evolving therapies. Int J Mol Sci. 2017;18(7):1338.
153. Chen R, Dawson DW, Pan S, Ottenhof NA, de Wilde RF, Wolfgang CL, et al. Proteins associated with pancreatic cancer survival in patients with resectable pancreatic ductal adenocarcinoma. Lab Investig. 2015;95(1):43–55.
154. Chandana SR, Woods LM, Maxwell F, Gandolfo R, Bekaii-Saab T. Risk factors for early-onset pancreatic ductal adenocarcinoma: a systematic literature review. Eur J Cancer. 2024;198:113471.
155. Hossain MS, Karuniawati H, Jairoun AA, Urbi Z, Ooi J, John A, et al. Colorectal cancer: a review of carcinogenesis, global epidemiology, current challenges, risk factors, preventive and treatment strategies. Cancers (Basel). 2022;14(7):1732.
156. Gui Y, Deng X, Li N, Zhao L. PRELP reduce cell stiffness and adhesion to promote the growth and metastasis of colorectal cancer cells by binding to integrin α5. Exp Cell Res. 2024;441(1):114151.
157. StatPearls. 2025.
158. Santos FP, O'Brien S. Small lymphocytic lymphoma and chronic lymphocytic leukemia: are they the same disease? Cancer J. 2012;18(5):396–403.
159. Iyer P, Wang L. Emerging therapies in CLL in the era of precision medicine. Cancers (Basel). 2023;15(5):1583.

160. Grover J, Chen XN, Korenberg JR, Recklies AD, Roughley PJ. The gene organization, chromosome location, and expression of a 55-kDa matrix protein (PRELP) of human articular cartilage. Genomics. 1996;38(2):109–17.
161. Masugi Y. The desmoplastic stroma of pancreatic cancer: multilayered levels of heterogeneity, clinical significance, and therapeutic opportunities. Cancers (Basel). 2022;14(13):3293.
162. Pandol S, Edderkaoui M, Gukovsky I, Lugea A, Gukovskaya A. Desmoplasia of pancreatic ductal adenocarcinoma. Clin Gastroenterol Hepatol. 2009;7(11 Suppl):S44–7.
163. Leiphrakpam PD, Patil PP, Remmers N, Swanson B, Grandgenett PM, Qiu F, et al. Role of keratan sulfate expression in human pancreatic cancer malignancy. Sci Rep. 2019;9(1):9665.
164. Kao WW, Liu CY. Roles of lumican and keratocan on corneal transparency. Glycoconj J. 2002;19(4–5):275–85.
165. Ping LY, Ishiwata T, Asano G. Lumican expression in alpha cells of islets in pancreas and pancreatic cancer cells. J Pathol. 2002;196(3):324–30.
166. López-Casas PP, López-Fernández LA. Gene-expression profiling in pancreatic cancer. Expert Rev Mol Diagn. 2010;10(5):591–601.
167. Deer EL, González-Hernández J, Coursen JD, Shea JE, Ngatia J, Scaife CL, et al. Phenotype and genotype of pancreatic cancer cell lines. Pancreas. 2010;39(4):425–35.
168. Hartupee C, Nagalo BM, Chabu CY, Tesfay MZ, Coleman-Barnett J, West JT, et al. Pancreatic cancer tumor microenvironment is a major therapeutic barrier and target. Front Immunol. 2024;15:1287459.
169. Lin W, Zhu X, Gao L, Mao M, Gao D, Huang Z. Osteomodulin positively regulates osteogenesis through interaction with BMP2. Cell Death Dis. 2021;12(2):147.
170. Jin L, Han B, Siegel E, Cui Y, Giuliano A, Cui X. Breast cancer lung metastasis: molecular biology and therapeutic implications. Cancer Biol Ther. 2018;19(10):858–68.
171. Chen YC, Sosnoski DM, Mastro AM. Breast cancer metastasis to the bone: mechanisms of bone loss. Breast Cancer Res. 2010;12(6):215.
172. Nuckhir M, Withey D, Cabral S, Harrison H, Clarke RB. State of the art modelling of the breast cancer metastatic microenvironment: where are we? J Mammary Gland Biol Neoplasia. 2024;29(1):14.
173. Alix-Panabières C, Vendrell JP, Slijper M, Pellé O, Barbotte E, Mercier G, et al. Full-length cytokeratin-19 is released by human tumor cells: a potential role in metastatic progression of breast cancer. Breast Cancer Res. 2009;11(3):R39.
174. Morrison C, Mancini S, Cipollone J, Kappelhoff R, Roskelley C, Overall C. Microarray and proteomic analysis of breast cancer cell and osteoblast co-cultures: role of osteoblast matrix metalloproteinase (MMP)-13 in bone metastasis. J Biol Chem. 2011;286(39):34271–85.
175. Taylor JA, Kuchel GA. Bladder cancer in the elderly: clinical outcomes, basic mechanisms, and future research direction. Nat Clin Pract Urol. 2009;6(3):135–44.
176. Matulewicz RS, Steinberg GD. Non-muscle-invasive bladder cancer: overview and contemporary treatment landscape of Neoadjuvant Chemoablative therapies. Rev Urol. 2020;22(2):43–51.
177. Martins-Lima C, Chianese U, Benedetti R, Altucci L, Jerónimo C, Correia MP. Tumor microenvironment and epithelial-mesenchymal transition in bladder cancer: cytokines in the game? Front Mol Biosci. 2022;9:1070383.
178. Rawla P, Barsouk A. Epidemiology of gastric cancer: global trends, risk factors and prevention. Prz Gastroenterol. 2019;14(1):26–38.
179. Hu B, El Hajj N, Sittler S, Lammert N, Barnes R, Meloni-Ehrig A. Gastric cancer: classification, histology and application of molecular pathology. J Gastrointest Oncol. 2012;3(3):251–61.
180. Sun C, Chen Y, Kim NH, Lowe S, Ma S, Zhou Z, et al. Identification and verification of potential biomarkers in gastric cancer by integrated Bioinformatic analysis. Front Genet. 2022;13:911740.
181. Johnson HJ, Rosenberg L, Choi HU, Garza S, Höök M, Neame PJ. Characterization of epiphycan, a small proteoglycan with a leucine-rich repeat core protein. J Biol Chem. 1997;272(30):18709–17.

182. Momenimovahed Z, Tiznobaik A, Taheri S, Salehiniya H. Ovarian cancer in the world: epidemiology and risk factors. Int J Women's Health. 2019;11:287–99.
183. Matulonis UA, Sood AK, Fallowfield L, Howitt BE, Sehouli J, Karlan BY. Ovarian cancer. Nat Rev Dis Primers. 2016;2:16061.
184. Ali AT, Al-Ani O, Al-Ani F. Epidemiology and risk factors for ovarian cancer. Prz Menopauzalny. 2023;22(2):93–104.
185. Deng L, Wang D, Chen S, Hu W, Zhang R. Epiphycan predicts poor outcomes and promotes metastasis in ovarian cancer. Front Oncol. 2021;11:653782.
186. He Y, Sun MM, Zhang GG, Yang J, Chen KS, Xu WW, et al. Targeting PI3K/Akt signal transduction for cancer therapy. Signal Transduct Target Ther. 2021;6(1):425.
187. Jackson JT, Mulazzani E, Nutt SL, Masters SL. The role of PLCγ2 in immunological disorders, cancer, and neurodegeneration. J Biol Chem. 2021;297(2):100905.
188. Yang Z, Li H, Hao J, Mei H, Qiu M, Wang H, et al. EPYC functions as a novel prognostic biomarker for pancreatic cancer. Sci Rep. 2024;14(1):719.
189. Jacks BE, Ekpemiro CU, Adeosun AA, Ogbonna UO, Ogundiran FT, Babalola F, et al. Molecular markers of pancreatic cancer: a 10-year retrospective review of molecular advances. Cureus. 2022;14(9):e29485.
190. Malshy K, Golijanin B, Khaleel S, Danaher K, Widener J, Schmit S, et al. Navigating management of localized prostate cancer in the geriatric population. Crit Rev Oncol Hematol. 2024;206:104600.
191. Zhao X, Hu D, Li J, Zhao G, Tang W, Cheng H. Database mining of genes of prognostic value for the prostate adenocarcinoma microenvironment using the cancer gene atlas. Biomed Res Int. 2020;2020:5019793.
192. Pattwell DM, Sheridan CM, Le Goff M, Bishop PN, Hiscott P. Localisation of opticin in human proliferative retinal disease. Exp Eye Res. 2010;90(3):461–4.
193. Hindson VJ, Gallagher JT, Halfter W, Bishop PN. Opticin binds to heparan and chondroitin sulfate proteoglycans. Invest Ophthalmol Vis Sci. 2005;46(12):4417–23.
194. Le Goff MM, Sutton MJ, Slevin M, Latif A, Humphries MJ, Bishop PN. Opticin exerts its anti-angiogenic activity by regulating extracellular matrix adhesiveness. J Biol Chem. 2012;287(33):28027–36.
195. Darwiche W, Gubler B, Marolleau JP, Ghamlouch H. Chronic lymphocytic leukemia B-cell Normal cellular counterpart: clues from a functional perspective. Front Immunol. 2018;9:683.
196. Mikaelsson E, Osterborg A, Tahmasebi Fard Z, Mahmoudi A, Mahmoudian J, Jeddi-Tehrani M, et al. Opticin, a small leucine-rich proteoglycan, is uniquely expressed and translocated to the nucleus of chronic lymphocytic leukemia cells. Exp Hematol Oncol. 2013;2(1):23.
197. Low SWY, Connor TB, Kassem IS, Costakos DM, Chaurasia SS. Small leucine-rich proteoglycans (SLRPs) in the retina. Int J Mol Sci. 2021;22(14):7293.
198. Madu CO, Wang S, Lu Y. Angiogenesis in breast cancer progression, diagnosis, and treatment. J Cancer. 2020;11(15):4474–94.
199. Le Goff MM, Lu H, Ugarte M, Henry S, Takanosu M, Mayne R, et al. The vitreous glycoprotein opticin inhibits preretinal neovascularization. Invest Ophthalmol Vis Sci. 2012;53(1):228–34.
200. Moreno-Perea M, Suárez-Castro A, Fraire-Soto I, Sifuentes-Padilla JL, Gutiérrez-Hernández R, Reyes-Estrada CA, et al. Proliferation, migration and invasion of breast cancer cell lines are inhibited by 1,5-Disubstituted Tetrazol-1,2,3-triazole hybrids through interaction with p53. Molecules. 2023;28(22):1.
201. Guillen KP, Fujita M, Butterfield AJ, Scherer SD, Bailey MH, Chu Z, et al. A human breast cancer-derived xenograft and organoid platform for drug discovery and precision oncology. Nat Cancer. 2022;3(2):232–50.
202. Nulali J, Zhan M, Zhang K, Tu P, Liu Y, Song H. Osteoglycin: an ECM factor regulating fibrosis and tumorigenesis. Biomol Ther. 2022;12(11):1674.
203. Qin W, Zhang J, Rong R, Zhang L, Gao H, Liu C, et al. Osteoglycin (OGN) promotes tumorigenesis of pancreatic cancer cell via targeting ID4. Tissue Cell. 2022;78:101867.

204. Dings MPG, Manoukian P, Waasdorp C, Quik JSE, Strijker M, Lodestijn SC, et al. Serum levels of iCAF-derived osteoglycin predict favorable outcome in pancreatic cancer. Int J Cancer. 2023;152(3):511–23.
205. Yu J, Mu Q, Fung M, Xu X, Zhu L, Ho RJY. Challenges and opportunities in metastatic breast cancer treatments: nano-drug combinations delivered preferentially to metastatic cells may enhance therapeutic response. Pharmacol Ther. 2022;236:108108.
206. Xu T, Zhang R, Dong M, Zhang Z, Li H, Zhan C, et al. Osteoglycin (OGN) inhibits cell proliferation and invasiveness in breast cancer via PI3K/Akt/mTOR signaling pathway. Onco Targets Ther. 2019;12:10639–50.
207. Wu Y, Sarkissyan M, Vadgama JV. Epithelial-mesenchymal transition and breast cancer. J Clin Med. 2016;5(2):13.
208. Rawla P, Sunkara T, Barsouk A. Epidemiology of colorectal cancer: incidence, mortality, survival, and risk factors. Prz Gastroenterol. 2019;14(2):89–103.
209. Hu X, Li YQ, Li QG, Ma YL, Peng JJ, Cai SJ. Osteoglycin (OGN) reverses epithelial to mesenchymal transition and invasiveness in colorectal cancer via EGFR/Akt pathway. J Exp Clin Cancer Res. 2018;37(1):41.
210. Hu X, Li YQ, Li QG, Ma YL, Peng JJ, Cai SJ. Osteoglycin-induced VEGF inhibition enhances T lymphocytes infiltrating in colorectal cancer. EBioMedicine. 2018;34:35–45.
211. Adams TS, Rogers LJ, Cuello MA. Cancer of the vagina: 2021 update. Int J Gynaecol Obstet. 2021;155(Suppl 1):19–27.
212. Bae SM, Min HJ, Ding GH, Kwak SY, Cho YL, Nam KH, et al. Protein expression profile using two-dimensional gel analysis in squamous cervical cancer patients. Cancer Res Treat. 2006;38(2):99–107.
213. Istiaq A, Ohta K. A review on Tsukushi: mammalian development, disorders, and therapy. J Cell Commun Signaling. 2022;16(4):505–13.
214. Dietel M, Bubendorf L, Dingemans AM, Dooms C, Elmberger G, García RC, et al. Diagnostic procedures for non-small-cell lung cancer (NSCLC): recommendations of the European expert group. Thorax. 2016;71(2):177–84.
215. Rostron BL, Wang J, Etemadi A, Thakur S, Chang JT, Bhandari D, et al. Associations between biomarkers of exposure and lung cancer risk among exclusive cigarette smokers in the Golestan cohort study. Int J Environ Res Public Health. 2021;18(14):7349.
216. Yamada T, Ohta K, Motooka Y, Fujino K, Kudoh S, Tenjin Y, et al. Significance of Tsukushi in lung cancer. Lung Cancer. 2019;131:104–11.
217. Huang H, Zhang D, Fu J, Zhao L, Li D, Sun H, et al. Tsukushi is a novel prognostic biomarker and correlates with tumor-infiltrating B cells in non-small cell lung cancer. Aging (Albany NY). 2021;13(3):4428–51.
218. Samant H, Amiri HS, Zibari GB. Addressing the worldwide hepatocellular carcinoma: epidemiology, prevention and management. J Gastrointest Oncol. 2021;12(Suppl 2):S361–S73.
219. Gosalia AJ, Martin P, Jones PD. Advances and future directions in the treatment of hepatocellular carcinoma. Gastroenterol Hepatol (N Y). 2017;13(7):398–410.
220. Lee YT, Sun N, Kim M, Wang JJ, Tran BV, Zhang RY, et al. Circulating tumor cell-based messenger RNA scoring system for prognostication of hepatocellular carcinoma: translating tissue-based messenger RNA profiling into a noninvasive setting. Liver Transpl. 2022;28(2):200–14.
221. Labgaa I, Villanueva A, Dormond O, Demartines N, Melloul E. The role of liquid biopsy in hepatocellular carcinoma prognostication. Cancers (Basel). 2021;13(4):659.
222. Mochida Y, Kaku M, Yoshida K, Katafuchi M, Atsawasuwan P, Yamauchi M. Podocan-like protein: a novel small leucine-rich repeat matrix protein in bone. Biochem Biophys Res Commun. 2011;410(2):333–8.
223. Noor H, Zaman A, Teo C, Sughrue ME. Methylation serves as a prognostic biomarker and associates with immune cell infiltration and immune checkpoint blockade response in lower-grade glioma. Int J Mol Sci. 2021;22(22):12572.
224. Kurman RJ, Shih IM. The origin and pathogenesis of epithelial ovarian cancer: a proposed unifying theory. Am J Surg Pathol. 2010;34(3):433–43.

225. Dion L, Mimoun C, Nyangoh Timoh K, Bendifallah S, Bricou A, Collinet P, et al. Ovarian cancer in the elderly: time to move towards a more logical approach to improve prognosis-a study from the FRANCOGYN group. J Clin Med. 2020;9(5):1339.
226. Teng C, Zheng H. Low expression of microRNA-1908 predicts a poor prognosis for patients with ovarian cancer. Oncol Lett. 2017;14(4):4277–81.
227. Zhang B, Cai FF, Zhong XY. An overview of biomarkers for the ovarian cancer diagnosis. Eur J Obstet Gynecol Reprod Biol. 2011;158(2):119–23.
228. Ren FJ, Cai XY, Yao Y, Fang GY. JunB: a paradigm for Jun family in immune response and cancer. Front Cell Infect Microbiol. 2023;13:1222265.
229. Chen R, Smith-Cohn M, Cohen AL, Colman H. Glioma subclassifications and their clinical significance. Neurotherapeutics. 2017;14(2):284–97.
230. Geng Y, Zuo P, Li XO, Zhang L. PODNL1 promotes cell proliferation and migration in glioma via regulating Akt/mTOR pathway. J Cancer. 2020;11(21):6234–42.
231. Shimizu-Hirota R, Sasamura H, Kuroda M, Kobayashi E, Saruta T. Functional characterization of podocan, a member of a new class in the small leucine-rich repeat protein family. FEBS Lett. 2004;563(1–3):69–74.
232. Burger M, Catto JW, Dalbagni G, Grossman HB, Herr H, Karakiewicz P, et al. Epidemiology and risk factors of urothelial bladder cancer. Eur Urol. 2013;63(2):234–41.
233. Mostafa MH, Sheweita SA, O'Connor PJ. Relationship between schistosomiasis and bladder cancer. Clin Microbiol Rev. 1999;12(1):97–111.
234. Santos LL, Santos J, Gouveia MJ, Bernardo C, Lopes C, Rinaldi G, et al. Urogenital schistosomiasis-history, pathogenesis, and bladder cancer. J Clin Med. 2021;10(2):205.
235. Liu X, He Y, Tong Y, Huang Q, Shang H, Chen Z, et al. PODNL1 promotes cell migration and regulates the epithelial/mesenchymal transition process in bladder cancer. Biochem Biophys Res Commun. 2022;620:165–72.
236. Eray A, Erkek-Özhan S. Classification of bladder cancer cell lines according to regulon activity. Turk J Biol. 2021;45(6):656–66.
237. Xue W, Yang L, Chen C, Ashrafizadeh M, Tian Y, Sun R. Wnt/β-catenin-driven EMT regulation in human cancers. Cell Mol Life Sci. 2024;81(1):79.
238. Muramaki M, Miyake H, Terakawa T, Kumano M, Sakai I, Fujisawa M. Expression profile of E-cadherin and N-cadherin in non-muscle-invasive bladder cancer as a novel predictor of intravesical recurrence following transurethral resection. Urol Oncol. 2012;30(2):161–6.
239. Nikitovic D, Aggelidakis J, Young MF, Iozzo RV, Karamanos NK, Tzanakakis GN. The biology of small leucine-rich proteoglycans in bone pathophysiology. J Biol Chem. 2012;287(41):33926–33.
240. Florencio-Silva R, Sasso GR, Sasso-Cerri E, Simões MJ, Cerri PS. Biology of bone tissue: structure, function, and factors that influence bone cells. Biomed Res Int. 2015;2015:421746.
241. Lin X, Patil S, Gao YG, Qian A. The bone extracellular matrix in bone formation and regeneration. Front Pharmacol. 2020;11:757.
242. Feng X. Chemical and biochemical basis of cell-bone matrix interaction in health and disease. Curr Chem Biol. 2009;3(2):189–96.
243. Ni GX, Li Z, Zhou YZ. The role of small leucine-rich proteoglycans in osteoarthritis pathogenesis. Osteoarthr Cartil. 2014;22(7):896–903.
244. Ameye L, Young MF. Mice deficient in small leucine-rich proteoglycans: novel in vivo models for osteoporosis, osteoarthritis, Ehlers-Danlos syndrome, muscular dystrophy, and corneal diseases. Glycobiology. 2002;12(9):107R–16R.
245. Dai G, Sun Y, Wei R, Xi L. Small leucine-rich proteoglycan. Curr Issues Mol Biol. 2023;45(7):6116–39.
246. Flannery CR. Usurped SLRPs: novel arthritis biomarkers exposed by catabolism of small leucine-rich proteoglycans? Arthritis Res Ther. 2006;8(2):106.
247. Merline R, Schaefer RM, Schaefer L. The matricellular functions of small leucine-rich proteoglycans (SLRPs). J Cell Commun Signaling. 2009;3(3–4):323–35.
248. Chen S, Birk DE. The regulatory roles of small leucine-rich proteoglycans in extracellular matrix assembly. FEBS J. 2013;280(10):2120–37.

249. Shainer R, Kram V, Kilts TM, Li L, Doyle AD, Shainer I, et al. Biglycan regulates bone development and regeneration. Front Physiol. 2023;14:1119368.
250. Hua R, Han Y, Ni Q, Fajardo RJ, Iozzo RV, Ahmed R, et al. Pivotal roles of biglycan and decorin in regulating bone mass, water retention, and bone toughness. Bone Res. 2025;13(1):2.
251. Bianco P, Fisher LW, Young MF, Termine JD, Robey PG. Expression and localization of the two small proteoglycans biglycan and decorin in developing human skeletal and non-skeletal tissues. J Histochem Cytochem. 1990;38(11):1549–63.
252. Ingram RT, Clarke BL, Fisher LW, Fitzpatrick LA. Distribution of noncollagenous proteins in the matrix of adult human bone: evidence of anatomic and functional heterogeneity. J Bone Miner Res. 1993;8(9):1019–29.
253. Han B, Li Q, Wang C, Chandrasekaran P, Zhou Y, Qin L, et al. Differentiated activities of decorin and biglycan in the progression of post-traumatic osteoarthritis. Osteoarthr Cartil. 2021;29(8):1181–92.
254. Markmann A, Hausser H, Schönherr E, Kresse H. Influence of decorin expression on transforming growth factor-beta-mediated collagen gel retraction and biglycan induction. Matrix Biol. 2000;19(7):631–6.
255. Heegaard AM, Xie Z, Young MF, Nielsen KL. Transforming growth factor beta stimulation of biglycan gene expression is potentially mediated by sp1 binding factors. J Cell Biochem. 2004;93(3):463–75.
256. Frey H, Schroeder N, Manon-Jensen T, Iozzo RV, Schaefer L. Biological interplay between proteoglycans and their innate immune receptors in inflammation. FEBS J. 2013;280(10):2165–79.
257. Merline R, Moreth K, Beckmann J, Nastase MV, Zeng-Brouwers J, Tralhão JG, et al. Signaling by the matrix proteoglycan decorin controls inflammation and cancer through PDCD4 and MicroRNA-21. Sci Signal. 2011;4(199):ra75.
258. Neill T, Schaefer L, Iozzo RV. Decorin: a guardian from the matrix. Am J Pathol. 2012;181(2):380–7.
259. Kram V, Shainer R, Jani P, Meester JAN, Loeys B, Young MF. Biglycan in the skeleton. J Histochem Cytochem. 2020;68(11):747–62.
260. Schaefer L, Babelova A, Kiss E, Hausser HJ, Baliova M, Krzyzankova M, et al. The matrix component biglycan is proinflammatory and signals through toll-like receptors 4 and 2 in macrophages. J Clin Invest. 2005;115(8):2223–33.
261. Chen XD, Fisher LW, Robey PG, Young MF. The small leucine-rich proteoglycan biglycan modulates BMP-4-induced osteoblast differentiation. FASEB J. 2004;18(9):948–58.
262. Mochida Y, Parisuthiman D, Yamauchi M. Biglycan is a positive modulator of BMP-2 induced osteoblast differentiation. Adv Exp Med Biol. 2006;585:101–13.
263. Wang X, Harimoto K, Xie S, Cheng H, Liu J, Wang Z. Matrix protein biglycan induces osteoblast differentiation through extracellular signal-regulated kinase and Smad pathways. Biol Pharm Bull. 2010;33(11):1891–7.
264. Berendsen AD, Fisher LW, Kilts TM, Owens RT, Robey PG, Gutkind JS, et al. Modulation of canonical Wnt signaling by the extracellular matrix component biglycan. Proc Natl Acad Sci USA. 2011;108(41):17022–7.
265. Raouf A, Ganss B, McMahon C, Vary C, Roughley PJ, Seth A. Lumican is a major proteoglycan component of the bone matrix. Matrix Biol. 2002;21(4):361–7.
266. Lee YS, Park SJ, Lee JY, Choi E, Kim BJ. Benefits of lumican on human bone health: clinical evidence using bone marrow aspirates. Korean J Intern Med. 2022;37(4):821–9.
267. Lee JY, Kim DA, Kim EY, Chang EJ, Park SJ, Kim BJ. Lumican inhibits Osteoclastogenesis and bone resorption by suppressing Akt activity. Int J Mol Sci. 2021;22(9):4717.
268. Wei S, Teitelbaum SL, Wang MW, Ross FP. Receptor activator of nuclear factor-kappa b ligand activates nuclear factor-kappa b in osteoclast precursors. Endocrinology. 2001;142(3):1290–5.
269. Wadhwa S, Embree MC, Kilts T, Young MF, Ameye LG. Accelerated osteoarthritis in the temporomandibular joint of biglycan/fibromodulin double-deficient mice. Osteoarthr Cartil. 2005;13(9):817–27.

270. Gill MR, Oldberg A, Reinholt FP. Fibromodulin-null murine knee joints display increased incidences of osteoarthritis and alterations in tissue biochemistry. Osteoarthr Cartil. 2002;10(10):751–7.
271. Murphy JM, Heinegård R, McIntosh A, Sterchi D, Barry FP. Distribution of cartilage molecules in the developing mouse joint. Matrix Biol. 1999;18(5):487–97.
272. Gori F, Schipani E, Demay MB. Fibromodulin is expressed by both chondrocytes and osteoblasts during fetal bone development. J Cell Biochem. 2001;82(1):46–57.
273. Hsueh MF, Önnerfjord P, Kraus VB. Biomarkers and proteomic analysis of osteoarthritis. Matrix Biol. 2014;39:56–66.
274. Zhao F, Bai Y, Xiang X, Pang X. The role of fibromodulin in inflammatory responses and diseases associated with inflammation. Front Immunol. 2023;14:1191787.
275. Couble ML, Bleicher F, Farges JC, Peyrol S, Lucchini M, Magloire H, et al. Immunodetection of osteoadherin in murine tooth extracellular matrices. Histochem Cell Biol. 2004;121(1):47–53.
276. Edsberg LE, Crowgey EL, Osborn PM, Wyffels JT. A survey of proteomic biomarkers for heterotopic ossification in blood serum. J Orthop Surg Res. 2017;12(1):69.
277. Rehn AP, Chalk AM, Wendel M. Differential regulation of osteoadherin (OSAD) by TGF-beta1 and BMP-2. Biochem Biophys Res Commun. 2006;349(3):1057–64.
278. Sommarin Y, Wendel M, Shen Z, Hellman U, Heinegârd D. Osteoadherin, a cell-binding keratan sulfate proteoglycan in bone, belongs to the family of leucine-rich repeat proteins of the extracellular matrix. J Biol Chem. 1998;273(27):16723–9.
279. Lin W, Gao L, Jiang W, Niu C, Yuan K, Hu X, et al. The role of osteomodulin on osteo/odontogenic differentiation in human dental pulp stem cells. BMC Oral Health. 2019;19(1):22.
280. Ramstad VE, Franzén A, Heinegård D, Wendel M, Reinholt FP. Ultrastructural distribution of osteoadherin in rat bone shows a pattern similar to that of bone sialoprotein. Calcif Tissue Int. 2003;72(1):57–64.
281. Arosarena O, Collins W. Comparison of BMP-2 and -4 for rat mandibular bone regeneration at various doses. Orthod Craniofac Res. 2005;8(4):267–76.
282. Johnson J, Shinomura T, Eberspaecher H, Pinero G, Decrombrugghe B, Höök M. Expression and localization of PG-Lb/epiphycan during mouse development. Dev Dyn. 1999;216(4–5):499–510.
283. Nuka S, Zhou W, Henry SP, Gendron CM, Schultz JB, Shinomura T, et al. Phenotypic characterization of epiphycan-deficient and epiphycan/biglycan double-deficient mice. Osteoarthr Cartil. 2010;18(1):88–96.
284. Aigner T, Fundel K, Saas J, Gebhard PM, Haag J, Weiss T, et al. Large-scale gene expression profiling reveals major pathogenetic pathways of cartilage degeneration in osteoarthritis. Arthritis Rheum. 2006;54(11):3533–44.
285. Funderburgh JL, Corpuz LM, Roth MR, Funderburgh ML, Tasheva ES, Conrad GW. Mimecan, the 25-kDa corneal keratan sulfate proteoglycan, is a product of the gene producing osteoglycin. J Biol Chem. 1997;272(44):28089–95.
286. Deckx S, Heymans S, Papageorgiou AP. The diverse functions of osteoglycin: a deceitful dwarf, or a master regulator of disease? FASEB J. 2016;30(8):2651–61.
287. Balakrishnan L, Nirujogi RS, Ahmad S, Bhattacharjee M, Manda SS, Renuse S, et al. Proteomic analysis of human osteoarthritis synovial fluid. Clin Proteomics. 2014;11(1):6.
288. Hamajima S, Hiratsuka K, Kiyama-Kishikawa M, Tagawa T, Kawahara M, Ohta M, et al. Effect of low-level laser irradiation on osteoglycin gene expression in osteoblasts. Lasers Med Sci. 2003;18(2):78–82.
289. Mizuno M, Fujisawa R, Kuboki Y. Bone chondroadherin promotes attachment of osteoblastic cells to solid-state substrates and shows affinity to collagen. Calcif Tissue Int. 1996;59(3):163–7.
290. Shen Z, Gantcheva S, Mânsson B, Heinegârd D, Sommarin Y. Chondroadherin expression changes in skeletal development. Biochem J. 1998;330(Pt 1):549–57.
291. Hessle L, Stordalen GA, Wenglén C, Petzold C, Tanner E, Brorson SH, et al. The skeletal phenotype of chondroadherin deficient mice. PLoS One. 2014;8(6):e63080.

292. Decker RS, Koyama E, Pacifici M. Articular cartilage: structural and developmental intricacies and questions. Curr Osteoporos Rep. 2015;13(6):407–14.
293. Raj PP. Intervertebral disc: anatomy-physiology-pathophysiology-treatment. Pain Pract. 2008;8(1):18–44.
294. Sophia Fox AJ, Bedi A, Rodeo SA. The basic science of articular cartilage: structure, composition, and function. Sports Health. 2009;1(6):461–8.
295. Dreischarf M, Shirazi-Adl A, Arjmand N, Rohlmann A, Schmidt H. Estimation of loads on human lumbar spine: a review of in vivo and computational model studies. J Biomech. 2016;49(6):833–45.
296. Arokoski JP, Jurvelin JS, Väätäinen U, Helminen HJ. Normal and pathological adaptations of articular cartilage to joint loading. Scand J Med Sci Sports. 2000;10(4):186–98.
297. Loeser RF, Goldring SR, Scanzello CR, Goldring MB. Osteoarthritis: a disease of the joint as an organ. Arthritis Rheum. 2012;64(6):1697–707.
298. Yao Q, Wu X, Tao C, Gong W, Chen M, Qu M, et al. Osteoarthritis: pathogenic signaling pathways and therapeutic targets. Signal Transduct Target Ther. 2023;8(1):56.
299. Caterson B, Flannery CR, Hughes CE, Little CB. Mechanisms involved in cartilage proteoglycan catabolism. Matrix Biol. 2000;19(4):333–44.
300. Alcaide-Ruggiero L, Cugat R, Domínguez JM. Proteoglycans in articular cartilage and their contribution to chondral injury and repair mechanisms. Int J Mol Sci. 2023;24(13):10824.
301. Chery DR, Han B, Zhou Y, Wang C, Adams SM, Chandrasekaran P, et al. Decorin regulates cartilage pericellular matrix micromechanobiology. Matrix Biol. 2021;96:1–17.
302. Li Q, Han B, Wang C, Tong W, Wei Y, Tseng WJ, et al. Mediation of cartilage matrix degeneration and fibrillation by Decorin in post-traumatic osteoarthritis. Arthritis Rheumatol. 2020;72(8):1266–77.
303. Chery DR, Han B, Li Q, Zhou Y, Heo SJ, Kwok B, et al. Early changes in cartilage pericellular matrix micromechanobiology portend the onset of post-traumatic osteoarthritis. Acta Biomater. 2020;111:267–78.
304. Vincent TL, McClurg O, Troeberg L. The extracellular matrix of articular cartilage controls the bioavailability of Pericellular matrix-bound growth factors to drive tissue homeostasis and repair. Int J Mol Sci. 2022;23(11):6003.
305. Johnstone B, Markopoulos M, Neame P, Caterson B. Identification and characterization of glycanated and non-glycanated forms of biglycan and decorin in the human intervertebral disc. Biochem J. 1993;292(Pt 3):661–6.
306. Koźma EM, Wisowski G, Kusz D, Olczyk K. The role of decorin and biglycan dermatan sulfate chain(s) in fibrosis-affected fascia. Glycobiology. 2011;21(10):1301–16.
307. Bayliss MT, Osborne D, Woodhouse S, Davidson C. Sulfation of chondroitin sulfate in human articular cartilage. The effect of age, topographical position, and zone of cartilage on tissue composition. J Biol Chem. 1999;274(22):15892–900.
308. Roughley PJ. Age-associated changes in cartilage matrix: implications for tissue repair. Clin Orthop Relat Res. 2001;391(Suppl):S153–60.
309. Oldberg A, Antonsson P, Moses J, Fransson LA. Amino-terminal deletions in the decorin core protein leads to the biosynthesis of proteoglycans with shorter glycosaminoglycan chains. FEBS Lett. 1996;386(1):29–32.
310. Barreto G, Soininen A, Ylinen P, Sandelin J, Konttinen YT, Nordström DC, et al. Soluble biglycan: a potential mediator of cartilage degradation in osteoarthritis. Arthritis Res Ther. 2015;17:379.
311. Avenoso A, D'Ascola A, Scuruchi M, Mandraffino G, Calatroni A, Saitta A, et al. The proteoglycan biglycan mediates inflammatory response by activating TLR-4 in human chondrocytes: inhibition by specific siRNA and high polymerized Hyaluronan. Arch Biochem Biophys. 2018;640:75–82.
312. Garantziotis S, Savani RC. Proteoglycans in toll-like receptor responses and innate immunity. Am J Physiol Cell Physiol. 2022;323(1):C202–C14.
313. Kondo S, Ishiguro N, Iwata H, Nakashima I, Isobe K. The effects of nitric oxide on chondrocytes and lymphocytes. Biochem Biophys Res Commun. 1993;197(3):1431–7.

314. Melrose J, Smith SM, Fuller ES, Young AA, Roughley PJ, Dart A, et al. Biglycan and fibromodulin fragmentation correlates with temporal and spatial annular remodelling in experimentally injured ovine intervertebral discs. Eur Spine J. 2007;16(12):2193–205.
315. Polgár A, Falus A, Koó E, Ujfalussy I, Seszták M, Szuts I, et al. Elevated levels of synovial fluid antibodies reactive with the small proteoglycans biglycan and decorin in patients with rheumatoid arthritis or other joint diseases. Rheumatology (Oxford). 2003;42(4):522–7.
316. Xu L, Li Z, Liu SY, Xu SY, Ni GX. Asporin and osteoarthritis. Osteoarthr Cartil. 2015;23(6):933–9.
317. Sakao K, Takahashi KA, Arai Y, Saito M, Honjyo K, Hiraoka N, et al. Asporin and transforming growth factor-beta gene expression in osteoblasts from subchondral bone and osteophytes in osteoarthritis. J Orthop Sci. 2009;14(6):738–47.
318. Leivonen SK, Lazaridis K, Decock J, Chantry A, Edwards DR, Kähäri VM. TGF-β-elicited induction of tissue inhibitor of metalloproteinases (TIMP)-3 expression in fibroblasts involves complex interplay between Smad3, p38α, and ERK1/2. PLoS One. 2013;8(2):e57474.
319. Li TF, O'Keefe RJ, Chen D. TGF-beta signaling in chondrocytes. Front Biosci. 2005;10:681–8.
320. Stanton H, Melrose J, Little CB, Fosang AJ. Proteoglycan degradation by the ADAMTS family of proteinases. Biochim Biophys Acta. 2011;1812(12):1616–29.
321. Sobhan MR, Mehdinejad M, Jamaladini MH, Mazaheri M, Zare-Shehneh M, Neamatzadeh H. Association between aspartic acid repeat polymorphism of the asporin gene and risk of knee osteoarthritis: a systematic review and meta-analysis. Acta Orthop Traumatol Turc. 2017;51(5):409–15.
322. Barreto G, Senturk B, Colombo L, Brück O, Neidenbach P, Salzmann G, et al. Lumican is upregulated in osteoarthritis and contributes to TLR4-induced pro-inflammatory activation of cartilage degradation and macrophage polarization. Osteoarthr Cartil. 2020;28(1):92–101.
323. Ritter SY, Collins J, Krastins B, Sarracino D, Lopez M, Losina E, et al. Mass spectrometry assays of plasma biomarkers to predict radiographic progression of knee osteoarthritis. Arthritis Res Ther. 2014;16(5):456.
324. Huang Z, Kraus VB. Does lipopolysaccharide-mediated inflammation have a role in OA? Nat Rev Rheumatol. 2016;12(2):123–9.
325. Brown S, Melrose J, Caterson B, Roughley P, Eisenstein SM, Roberts S. A comparative evaluation of the small leucine-rich proteoglycans of pathological human intervertebral discs. Eur Spine J. 2012;21(Suppl 2):S154–9.
326. Noyori K, Jasin HE. Inhibition of human fibroblast adhesion by cartilage surface proteoglycans. Arthritis Rheum. 1994;37(11):1656–63.
327. Sjöberg A, Onnerfjord P, Mörgelin M, Heinegård D, Blom AM. The extracellular matrix and inflammation: fibromodulin activates the classical pathway of complement by directly binding C1q. J Biol Chem. 2005;280(37):32301–8.
328. Son M. Understanding the contextual functions of C1q and LAIR-1 and their applications. Exp Mol Med. 2022;54(5):567–72.
329. Melrose J, Fuller ES, Roughley PJ, Smith MM, Kerr B, Hughes CE, et al. Fragmentation of decorin, biglycan, lumican and keratocan is elevated in degenerate human meniscus, knee and hip articular cartilages compared with age-matched macroscopically normal and control tissues. Arthritis Res Ther. 2008;10(4):R79.
330. Shu CC, Flannery CR, Little CB, Melrose J. Catabolism of Fibromodulin in developmental rudiment and pathologic articular cartilage demonstrates novel roles for MMP-13 and ADAMTS-4 in C-terminal processing of SLRPs. Int J Mol Sci. 2019;20(3):579.
331. Tessier S, Tran VA, Ottone OK, Novais EJ, Doolittle A, DiMuzio MJ, et al. TonEBP-deficiency accelerates intervertebral disc degeneration underscored by matrix remodeling, cytoskeletal rearrangements, and changes in proinflammatory gene expression. Matrix Biol. 2020;87:94–111.
332. Johnson ZI, Shapiro IM, Risbud MV. Extracellular osmolarity regulates matrix homeostasis in the intervertebral disc and articular cartilage: evolving role of TonEBP. Matrix Biol. 2014;40:10–6.

333. Zappia J, Tong Q, Van der Cruyssen R, Cornelis FMF, Lambert C, Pinto Coelho T, et al. Osteomodulin downregulation is associated with osteoarthritis development. Bone Res. 2023;11(1):49.
334. Kamekura S, Hoshi K, Shimoaka T, Chung U, Chikuda H, Yamada T, et al. Osteoarthritis development in novel experimental mouse models induced by knee joint instability. Osteoarthr Cartil. 2005;13(7):632–41.
335. Bergen DJM, Tong Q, Shukla A, Newham E, Zethof J, Lundberg M, et al. Regenerating zebrafish scales express a subset of evolutionary conserved genes involved in human skeletal disease. BMC Biol. 2022;20(1):21.
336. Tashima T, Nagatoishi S, Sagara H, Ohnuma S, Tsumoto K. Osteomodulin regulates diameter and alters shape of collagen fibrils. Biochem Biophys Res Commun. 2015;463(3):292–6.
337. Zhao W, von Kroge S, Jadzic J, Milovanovic P, Sihota P, Luther J, et al. Osteomodulin deficiency in mice causes a specific reduction of transversal cortical bone size. J Bone Miner Res. 2024;39(7):1025–41.
338. Rajasekaran S, Tangavel C, Soundararajan DCR, Nayagam SM, Matchado MS, Muthurajan R, et al. Proteomic signatures of healthy intervertebral discs from organ donors: a comparison with previous studies on discs from scoliosis, animals, and trauma. Neurospine. 2020;17(2):426–42.
339. Novais EJ, Narayanan R, Canseco JA, van de Wetering K, Kepler CK, Hilibrand AS, et al. A new perspective on intervertebral disc calcification-from bench to bedside. Bone Res. 2024;12(1):3.
340. Deere M, Johnson J, Garza S, Harrison WR, Yoon SJ, Elder FF, et al. Characterization of human DSPG3, a small dermatan sulfate proteoglycan. Genomics. 1996;38(3):399–404.
341. Farrán A, Valverde-Franco G, Tío L, Lussier B, Fahmi H, Pelletier JP, et al. In vivo effect of opticin deficiency in cartilage in a surgically induced mouse model of osteoarthritis. Sci Rep. 2018;8(1):457.
342. Bentz H, Nathan RM, Rosen DM, Armstrong RM, Thompson AY, Segarini PR, et al. Purification and characterization of a unique osteoinductive factor from bovine bone. J Biol Chem. 1989;264(34):20805–10.
343. Shintani N, Siebenrock KA, Hunziker EB. TGF-ß1 enhances the BMP-2-induced chondrogenesis of bovine synovial explants and arrests downstream differentiation at an early stage of hypertrophy. PLoS One. 2013;8(1):e53086.
344. Bentz H, Thompson AY, Armstrong R, Chang RJ, Piez KA, Rosen DM. Transforming growth factor-beta 2 enhances the osteoinductive activity of a bovine bone-derived fraction containing bone morphogenetic protein-2 and 3. Matrix. 1991;11(4):269–75.
345. Shukunami C, Ishizeki K, Atsumi T, Ohta Y, Suzuki F, Hiraki Y. Cellular hypertrophy and calcification of embryonal carcinoma-derived chondrogenic cell line ATDC5 in vitro. J Bone Miner Res. 1997;12(8):1174–88.
346. Haglund L, Tillgren V, Addis L, Wenglén C, Recklies A, Heinegård D. Identification and characterization of the integrin alpha2beta1 binding motif in chondroadherin mediating cell attachment. J Biol Chem. 2011;286(5):3925–34.
347. Bisson Daniel G, Lama P, Abduljabbar F, Rosenzweig DH, Saran N, Ouellet JA, et al. Facet joint degeneration in adolescent idiopathic scoliosis. JOR Spine. 2018;1(2):e1016.
348. Haglund L, Tillgren V, Önnerfjord P, Heinegård D. The C-terminal peptide of chondroadherin modulates cellular activity by selectively binding to heparan sulfate chains. J Biol Chem. 2013;288(2):995–1008.
349. Batista MA, Nia HT, Önnerfjord P, Cox KA, Ortiz C, Grodzinsky AJ, et al. Nanomechanical phenotype of chondroadherin-null murine articular cartilage. Matrix Biol. 2014;38:84–90.
350. Paracuellos P, Kalamajski S, Bonna A, Bihan D, Farndale RW, Hohenester E. Structural and functional analysis of two small leucine-rich repeat proteoglycans, fibromodulin and chondroadherin. Matrix Biol. 2017;63:106–16.

Regulatory Role and Significance of Class I SLRPs: Biglycan, Decorin, and Asporin

3

Bibhushan Dhungyal and Pramila Koirala

Abstract

The main purpose of this chapter is to describe and discuss the role of class I SLRPs biglycan, decorin, and asporin in the diseases of the matrix to reveal their essential regulatory roles. Class I SLPRs are expressed during the development of cartilage, bones, tendons, and ligaments, including collagen fibrillogenesis, cell differentiation, migration, inflammation, angiogenesis, and even cancer. Notably biglycan and decorin are considered as damage-associated molecular pattern recognition molecules influencing the regulation of signaling pathways that contribute to tissue pathogenesis; thus, this chapter highlights the functional significance and role of class I SLRPs.

Keywords

Biglycan · Decorin · Asporin · Regulation · Disease · Cells

3.1 Class I Small Leucine-Rich Proteoglycans (SLRPs)

They are a group of small proteoglycans that includes biglycan (BGN), decorin (DCN), and asporin (ASPN), essential for regulating extracellular matrix assembly as well as cell signaling. Structurally, DCN and BGN contain chondroitin or

B. Dhungyal (✉)
Department of Physiology, Sikkim Manipal Institute of Medical Sciences, Sikkim Manipal University, Tadong, Sikkim, India
e-mail: Bibhushan.d@smims.smu.edu.in

P. Koirala
Department of Microbiology, Nar Bahadur Bhandari Government College, Sikkim University, Tadong, Sikkim, India

P. Lama (ed.), *The Biological Role of Small Leucine-Rich Proteoglycans (SLRPs) in Matrix Homeostasis and Diseases*,
https://doi.org/10.1007/978-981-96-7890-7_3

dermatan sulfate side chains while ASPN does not [1, 2] show them, and instead they contain a stretch of asporin residues, an acidic domain positioned in either the C or N terminal region. Biglycan and decorin have been found in a variety of skeletal regions, including periosteum, vascular canals, subchondral bone region, articular, and epiphyseal cartilage [3]. ASPN is associated with the formation of cartilage and the development of tendons and ligaments [4]. Class I SLRP are characterized by their highly conserved intron/exon junctions and a consistent structure comprising of eight exons with a similar organizational pattern [5].

Recent research on class I SLPRs—BGN, DCN, and ASPN—reveales an essential regulatory role [6, 7] in extracellular matrix assembly, cell maturation, differentiation, and migration, including cell signaling, inflammation-associated neo-angiogenesis, and invasion of cancer cells as some good examples [6, 8]. Numerous pathogenic mechanisms causing osteoporosis, osteosarcoma, cardiovascular disease, cancer, and many other disease conditions have been linked to SLRP genes [9, 10] and they are known to regulate cellular processes; therefore, any deviation in its structural integrity and/or biological function are detrimental in tissue pathogenesis, further making the progression of diseases inevitable [5]. Thus, the main aim of this chapter is to describe and compare these changes with their functional relevance.

3.1.1 Biglycan

Biglycan is a class I SLRP expressed in almost all the tissues within the extracellular matrix. It comprises of a 42 kDa protein core with eleven leucine-rich repeats (LLRs) and two covalently bound chondroitin sulfate or dermatan sulfate glycosaminoglycans side chains at the N-terminal. BGN interacts with type I, II, III, and VI collagen fiber or elastin and thus improves collagen fibrillogenesis, which in turn affects the composition and mechanical strength of the bone [11] and plays pivotal role in organizing the assembly of collagen-I fibers. BGN deficiency or targeted disruption of the BGN gene results in abnormal collagen fibril morphology and an osteoporosis-like phenotype characterized by low bone mass [12]. BGN is known to function as a signaling molecule, and its impact on cell proliferation varies depending on cell types in the tissue region where it is present. Further, BGN can alter several cell-signaling pathways and activities by binding to the various growth factors such as TGF-β, cytokine TNF-α, and chemokine molecules. It is a known ligand for members of the selectin, cadherin, and integrin families [13]. BGN has been reported to possess many therapeutic potentials influencing the progression and recovery of various diseases and disorders, and plays a significant role in stabilizing dystrophin-associated protein complex (DAPC), inducing phosphorylation of muscle-specific tyrosine kinase, thereby preventing muscular dystrophy [11, 14].

3.1.2 Expression of Biglycan

BGN is known to be expressed in many tissues, including bone, cartilage, blood vessels, muscles, and tendons. It is vital for postnatal skeletal growth and is involved in blood vessel formation and tissue homeostasis. In both regenerative and dystrophic muscle, BGN is upregulated. Developing tendon also express significant levels of BGN, and it has been reported that in human fetal muscle the expression of BGN transcript reaches its highest level around 21 weeks of gestation [14]. Various studies have revealed that BGN is overexpressed in the malignant mesenchyme of endometrial cancer. Furthermore, a substantial correlation between BGN expression and the histological grade of endometrial cancer is reported, and the International Federation of Gynecology and Obstetrics suggest BGN act as a potential prognosis marker in endometrial cancer [15]. BGN overexpression is documented in several types of cancer, including gastric, colorectal, odontogenic, esophageal squamous cell carcinoma, and intrahepatic cholangiocarcinoma, and this changes occasionally corresponds with advancement of the disease [16, 17].

3.1.3 Role of Biglycan in Bone Formation

BGN is known to regulate bone homeostasis, as well as remodeling and mineralization of skeletal tissues [18]. BGN stimulates the BMP/TGFβ pathways, which leads to the transcription of osteoblast-related genes, causing osteoblast differentiation. Furthermore, BGN enhances the efficacy of the Wnt/β-catenin signaling pathway by binding to the Wnt3a ligand and its LRP6 receptor, which also aids in osteoblast differentiation. BGN-deficient mice displays an osteoporosis-like phenotype/osteopenia, suggesting a regulatory role in bone formation [19, 20]. Various ex vivo model studies have revealed that age-related loss of GAGs is linked to a reduction in BGN and a decline in bone toughness. Recent in vivo studies have shown BGN deficiency induces reduction of total glycosaminoglycans in bone matrix and decreases amount of bound water impacting bone toughness in BGN knockout mice [21]. BGN is also reported to modulate angiogenesis and promote bone formation during fracture healing, by regulating the synthesis of bone morphogenetic protein-2 (BMP2) [22].

3.1.4 Role of Biglycan in Muscular Dystrophies

BGN is expressed in muscle tissues, and past studies indicate BGN binds to dystroglycan, and their expression is increased in the muscle tissue of mdx mice, a model for Duchenne muscular dystrophy (DMD). Thereafter, it was also discovered that DMD patient exhibit higher levels of BGN [23]. Both dystrophic and regenerative muscle with elevated levels of BGN presents extracellular ligand for dystrophin-associated protein complex (DAPC). Additionally, BGN binds to collagen VI, and mutations in this site result in congenital muscular dystrophy, which is typified by abnormalities of the connective tissue. Past studies suggest significance of BGN in muscle and connective tissue development, and further, biglycan plays a role in the pathogenesis of collagen VI-associated congenital muscular dystrophies [24, 25]. During the process of skeletal muscle development, BGN controls the availability of TGF-β as BGN lowers TGF-β signalling and permits myogenesis by binding to another TGF-β molecule and sequestering it from the membrane of myotubes [26].

3.1.5 Role of Biglycan in Modulation of Inflammatory Process

Soluble BGN, when released from ECM, functions as a warning signal that indicates tissue stress or injury, which triggers the innate and adaptive immune systems by binding to Toll-like receptors (TLRs) 2 and 4 on the immune cells [11]. BGN binds to the coreceptor CD-14 on the surface of macrophages triggering the production of TNF-α and CCL2 via TLR2/4 receptor, HSP70 thorough TLR2, and CCL5 by engaging the TLR4 receptor. This signal activates cytokines and chemokines, which results in the accumulation of M1 macrophages, thereby further accelerating the inflammatory responses. Recent studies have demonstrated the role of BGN in controlling T cell biology through overexpression or genetic ablation in mice with lupus nephritis (LN). BGN triggers the synthesis and secretion of proinflammatory molecules IL-1β and IL-18 through TLR2/4 and $P2X_4/P2X_7$ receptors. Furthermore, the binding of TLR 2/4 with $P2X_4/P2X_7$ receptors causes the BGN to induce the production of inflammasome, a signaling complex that can induce inflammation in response to pathogen-associated molecular patterns (PAMPs) and/or damage-associated molecular patterns (DAMPs) recognition molecule. This pathway finally leads to the activation and production of caspase-1 and potentially degradative forms of circulating cytokines [27]. Various studies have revealed a reduction of inflammasome activation in BGN-deficient mouse models of LN and unilateral ureteral obstruction (UUO), initating sterile inflammation and fibrosis thus suggesting the potential role of BGN in renal inflammation [11].

3.2 Decorin

Decorin (DCN) is a vital component of the connective tissue and the ECM, and they are known to regulate numerous cellular growth factors, ECM assembly, and tumor suppression. The core protein of DCN has eleven repeating leucine-rich motifs and a GAG side chains covalently attached to the amino terminus via the serine residue [28]. GAG side chain of DCN is usually presented as chondroitin sulfate (CS) or dermatan sulfate (DS), depending on the tissue type [29]. Thus, bone (B) and cartilage mainly have CS side chain, while skin, tendons, and intima of arteries show dermatan sulphate (DS) chains [30]. DCN is also synthesized by dermal fibroblasts and epidermal cells that bind to collagens [28] and operates complex signalling pathways that involve expression of growth factors, proteases, receptor tyrosine kinases, and other ECM components [28, 31]. Therefore, past research indicates its vital function in tumor angiogenesis [32], metastasis [33], autophagy [34], cell cycle [35, 36], wound repair [37], inflammatory diseases, and immune regulation [30].

3.2.1 Expression of Decorin

Dynamic balance of the extracellular matrix is maintained by DCN molecules, Fig. 3.1 mainly in fibrous connective tissue and in collagen fiber synthesis of cartilage, tendons, dermis, and cornea [29]. DCN is mainly expressed by fibroblasts, endothelial cells, and epithelial cells [30, 38], modulating transforming growth factor-beta (TGF-β) [39], interleukin-1 (IL-1) [40], tumor necrosis factor-alpha (TNF-α) [41], and even mechanical stress [42].

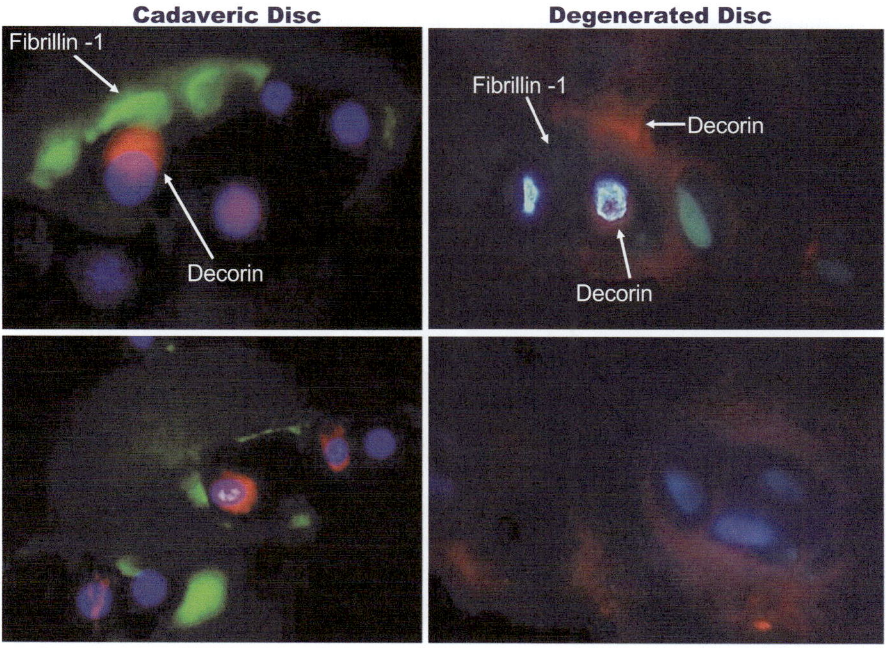

Fig. 3.1 Comparisons between a non-degenerated control cadaveric disc and a degenerated disc with double immunostaining techniques. Monoclonal primary antibodies were used to detect decorin and fibrillin-1. Fibrillin-1 (green) and decorin (red) show well-maintained organization in the non-degenerated cadaveric control discs, while this distinction is lost in the degenerated disc tissue, with diffused staining for the molecules. Blue indicates cell nuclei stained with DAPI. Scale bar 50um. (Figure adapted from Dr. Polly Lama, Molecular Matrix Lab)

3.2.2 Role of Decorin in Inflammatory Diseases and Immune Regulation

DCN is expressed during inflammation-related angiogenesis and in proinflammatory responses. IL-6 and IL-10 upregulate DCN in unitary smooth muscle fibers and endothelial cells [30] while on the other extreme, tumor necrosis factor-α (TNF-α) and TGF-β downregulate the transcription of DCN [43]. Studies on expression pattern of IL-1 and IL-4 with decorin have controversial findings, with both positive and negative effect on its transcription [43–45]. Immune cells like peritoneal macrophages, CD4+T and CD8+T cells are reported to upregulate expression of decorin, and they bind to macrophage surface receptor TLR-2 and TLR-4, triggering an acute activation of P38, NF-κB, and ERK1/2 inflammatory pathways, which leads to the synthesis of proinflammatory factors IL-12p70, TNF-α, and IL-10 [30]. DCN also act as a transcription inducer of tumor suppressor programmed cell death 4 (PDCD4) mediated through TLR2/4 signaling. PDCD4 is a translational suppressor; hence, through this pathway, proinflammatory environment of the matrix is maintained [46]. In several disease condition, DCN plays proinflammatory role, and the macrophage inflammatory protein-1 (MCP-1) level is increased in chronic pancreatitis [47]. DCN is also known to enhance the activity of interferon-γ (IFN-γ) facilitating syntheis of signal transduction transcriptional activator-1 (STAT-1) [48] complex. Additionally, it has been shown that only intact DCN protein core with glycosaminoglycan (GAG) chain is biologically active, and capable of initiating proinflammatory signals. Enzymes like matrix metalloproteinases (MMPs) 2, 3, 7, and 8 denature/breakdown the various matrix component, resulting in the release of DCN-bound cytokines, TGF-β or TNF-α. Therefore, the expression and regulation of DCN in the inflammatory condition are complex [32, 49, 50].

3.2.3 Role of Decorin in Tumorigenesis

DCN has been described as "the guardian from the matrix" because of its profound action to inhibit tumorigenesis [36]. It is well-known that genesis and progression of tumor is not just the result of DNA mutation but is also depend on specific malignant cells that subsist the tumor microenvironment [51] mainly formed by cancer-associated fibroblasts, nonmalignant cells, inflammatory cells, various types of immune cells, tissue macrophages, and angiogenic vascular system. Tumor microenvironment includes extracellular matrix components that play an important role in modulating cellular signalling [52], Fig. 3.1. DCN and BGN regulate tumor microenvironment and are known to modulate tumor-associated pathways involving autophagy, apoptosis, angiogenesis, proliferation, and inflammation. Diverse functions of DCN are because of its ability to bind with various components of the cellular receptors and ECM that initiates intracellular signalling cascades [53]. DCN also act as a transmembrane protein RTK inhibitor that regulates mesenchymal-epithelial transition factor (Met) receptor, epidermal growth factor receptor (EGFR), and vascular endothelial growth factor receptor-2 (VEGFR-2). RTKs are internalized and denatured, thus limiting angiogenesis and inhibiting cancer progression [54].

Tumorigenesis is dependent on angiogenesis because vascularization is essential for tumor survival and progression. In a nonmalignant state, DCN promotes angiogenesis, especially by supporting migration and adhesion of endothelial cells [55]. DCN is also known to preserve endothelial cells from hypoglycemia by facilitating angiogenesis via IGF-1R (insulin-like growth factor type-1 receptor) signalling. Interestingly, a number of studies suggest angiostatic effect of DCN suppresses tumor progression [54]. DCN brings about angiogenesis by several different mechanisms involving the Met signalling pathway, promoting the activities of TIMP-3, while Met-induced VEGFA (vascular endothelial growth factor A) expression and hypoxia-inducible factor-1α (HIF-1α) are suppressed. Further, lack of HIF-1α inhibits Met expressions [55, 56] involving a mechanism where DCN interacts with VEGFR-2, a surface receptor of endothelial cells that supports vascularization [32, 57], Thus, in malignant tissue, DCN acts as an antagonist of VEGFR-2 [41]. In addition, several past studies explored DCN-mediated angiostasis through DCN-EGFR interactions, and found Rho-associated coiled kinase-1 (ROCK1) and Rho (RAS homolog family member A) pathways were supressed and decorin exerts an antiangiogenic effects as tissue inhibitor of metalloproteinases-3 (TIMP-3) and thrombospondin-1. Thus, DCN promotes antitumor activities through a wide range of pathways that are reported to initiate autophagic complexes including inactivation of autophagy inhibitors, and activation of AMP-activated protein kinase (AMPK). This interactions also initiate pathways that inhibits the mTOR (mammalian target of rapamycin) complex while stimulating VPS34 (vacuolar protein sorting 34), thus contributing toward autophagy [30, 55, 58].

3.2.4 Decorin and Antifibrosis

DCN is known to inhibit fibrosis and promote scar tissue formation contributing towards healing [59] in the extracellular matrix [60], including inhibition of cell-cycle regulating cell proliferation [35, 36] and by controling fibrillogenesis [61]. There are reports of potential wound healing in various disease conditions, including cancer [37]. Interestingly, DCN is also found to be involved in the regulation of corneal transparency. Thus, visual abnormalities and corneal opacity linked to congenital stromal dystrophy is caused by mutations in the DCN gene [29] facilitating corneal wound healing by antagonizing the regulatory cytokine-TGFβ. Overexpression of TGF-β in corneal injury is known to induce keratocyte transdifferentiation to myofibroblasts resulting in fibrosis of the cornea [29, 37].

3.2.5 Role of Decorin in Autophagy

Recent research reveals role of DCN in the regulation of autophagy, a homeostatic mechanism by which cytoplasmic contents are degraded and recycled. DCN interacts directly with receptor tyrosine kinases, such as vascular endothelial growth factor receptor-2 (VEGFR-2) and mesenchymal epithelial transition factor receptor (Met), to induce autophagy in endothelial cells and mitophagy in breast cancer cells [58]. Paternally expressed gene-3 (Peg-3) is DCN-inducible, and genomically imprinted tumor suppressor that contributes to this process. Further, upregulation of kinase Vps34, a class III phosphoinositide kinase, is necessary for Peg3 activation. Additionally, DCN cause the inhibition of Bcl-2/Beclin 1 complexes to dissolve the differential development of Vps34/Beclin-1, thereby activating pro-autophagic AMPK-mediated signalling cascades while suppressing Akt/mTOR/p70S6K activity to prevent anti-autophagic action [62]. In vitro and in vivo expression of DCN at the mRNA and protein levels is shown to be effected by nutritional deprivation, and this process controls the transcriptional level blocking by the conventional mammalian target of rapamycin (mTOR) pathway, resulting in an autophagic response [34]. A few studies have also indicated that autophagy plays a key role in tumor growth prevention in early stages by protecting the healthy cells from damage [63].

3.2.6 Role of Decorin in Wound Repair

DCN has been found to be expressed mostly in the reticular dermis of the skin, and it is also a type I collagen-binding proteoglycan which a regulatory role in the formation of the collagen fibril structure in the skin. Thus, DCN is known to play an important role in tissue repair and wound healing [64], and it affects the cellular phenotype and tensile strength of the tissue [37]. DCN averts tissue fibrosis and promotes regeneration by regulatory mechanisms including TGF-β1 neutralizing or inhibitory activity. In vivo studies reveals that as compared to the nontargeted DCN, the fused DCN-targeting peptide gets deposited in the wounds thereby promoting wound healing and suppressing scar formation [65]. These findings highlight the biotherapeutic potential of decorin (DCN) in managing pathological conditions by reducing or preventing excessive scar formation. DCN may offer promising strategies for treating fibrosis-related disorders and improving tissue repair outcomes. Its ability to modulate extracellular matrix dynamics and inhibit key processes like the cell cycle and fibrillogenesis positions it as a valuable candidate for therapeutic development.

3.3 Asporin

Asporin (ASPN) is the third member of the class I SLRPs first identified in 2001, and it was also the first member of the LLR family of proteins to be discovered [66, 67], where it was purified from human articular cartilage and meniscus. ASPN name reflects a unique aspartic acid repeat at the N-terminus and confirms an overall similarity to DCN (50% amino acid sequence similarity) [1, 66, 68]. Although closely related to DCN and BGN, it is not a true proteoglycan as it lacks a glycosaminoglycan attachment site and contains an aspartic acid (D) repeat at its amino terminus of mature protein [68]. Number of consecutive aspartate residues varies from eleven to fifteen indicating polymorphism [69], and it is encoded by protein that consists of 380 amino acids, containing a pro-peptide with eighteen amino acid and putative signal peptide with fourteen amino acid, four N-cysteine terminals (C-X3-C-X-C-X6-C), two C-terminal cysteines, a putative pro-peptide, along with ten leucine-rich repeats [70, 71].

3.3.1 Expression of Asporin

The ASPN gene in humans has eight exons that span 26 kilobases pair on chromosome 9q31.1-32 [66], and their mRNA is expressed in almost all the tissues with greater density in liver, uterus, cardiac muscle, and aorta and in mineralizing tissues including bone, periodontal ligament, and osteoarthritic articular cartilage [68, 72]. Prenatal mRNA expression of asporin in mouse embryos shows it is absent in both articular and growth plate cartilage but are expressed in long bones, some of the flat bones, including vertebrae, mandible, and perichondrium of the maxilla [1, 70]. ASPN expression in the knee joints of 8-week-old mice is found predominantly in the tendon and periosteum and they are absent in articular cartilage and growth plate [70]. ASPN is expressed in the periodontal ligament (PDL) and are identified as PLAP-1 (periodontal ligament-associated protein-1). PDL is lodged in the inner wall of the tooth supporting the bone socket that links teeth to the alveolar bone proper, and are involved in periodontal tissue homeostasis, supporting nutrition, and initiates apoptotic repair [73]. ASPN expression in healthier tissue occurs at the boundary separating the growth plate from the articular cartilage [4]. ASPN is found to be upregulated in various pathological processes, and are identified in diseased cardiac myocytes [74], fibrotic lung tissue [75], degenerated articular cartilage [76], and keloid [77] disorder.

3.3.2 Role of Asporin in Cancer Cell Progression

Recent reports suggest asporin plays a negative and positive roles in the development and progression of cancer, and mainly controls tumor progression in breast, pancreas, gastric gland, and colon [71]. It is also known to suppress triple-negative breast cancer via numerous complex signalling cascades, and ASPN regulates three important cancer-related pathways, namely, TGF-beta, EGFR, and CD44 [71, 78].

3.3.2.1 TGF-Beta Signaling

ASPN promotes cancer cell proliferation by binding with CD44 and Smad 2/3 by promoting phosphorylation of EGFR. On the other hand, it inhibits cancer cell invasion and metastasis by binding to TGF-β in the extracellular matrix [71, 79]. Cancer cell properties and behavior are largely influenced by TGF-β/Smad2/3 signalling pathways that operate through unique TGF-β serine-threonine kinases either promoting or suppressing tumor progression [80]. TGF-β secreted by tumor cells is known to modulate tumor cell microenvironment through a paracrine and autocrine fashion [81, 82]. TGF-β1exhibit a dual role in breast cancer cell progression. In the early stage, it mediates growth arrest and cancer cell apoptosis, but in the later stages, it promotes the malignancy of breast cancer cells [83]. In the stroma of breast cancer lesion, ASPN density is found to be remarkably high, suggesting upregulation supporting tumor progression. It was also reported that ASPN expression is facilitated by TGF-β1 and inhibited by IL-1β in cancerous fibroblast cells [71]. In osteoarthritis, ASPN is known to interact directly with TGF-β1, thereby inhibiting downstream gene expression of collagen and aggrecan [84]. ASPN is also known to interact with Smad2/3 that leads to the entry of p-Smad2/3 into the nucleus, inducing colorectal cancer progression via EMT, and this was the first evidence to prove that ASPN is not an extracellular matrix component in cancer but it rather acts as an intracellular molecule [71]. Altogether, the TGF-β signaling pathway tend to be modulated by ASPN through extracellular TGF-β1 or cytoplasmic Smad2/3, and under certain conditions, ASPN can acts as a tumor suppressor or may aid tumor progression.

3.3.2.2 Epidermal Growth Factor Receptor (EGFR) Signaling Pathways

EGFR, is a receptor tyrosine kinase (RTKs), that plays an important role in cancer cell proliferation or apoptosis; and this modulation acts via AKT/PI3K/mTOR and RAS/RAF/MEK/ERK pathways [85]; therefore, a number of studies have indicated that EGFR is a potential target for cancer therapy. Poor prognosis of gastric cancer is always linked to upregulation of EGFR. Recent studies suggest that ASPN facilitates the activation of EGFR and downstream ERK1/2 pathways [86–88]. ASPN present in the cytoplasm and nucleus of gastric cancer cell lines is found to interact with phosphatidylserine decarboxylase proenzyme-2 (PSMD-2) to enhance degradation of tumor suppressor factors (PTEN, WIP1, and DUSP7). This leads to the activation of P38/MAPK, PI3K/AKT, and MAPK/ERK signalling pathways [89]. ASPN is reported to facilitate colorectal cancer cell proliferation by activating phospho-EGF receptor, phospho-Src family receptor (p-SrcT^{yr416}), and phospho-cortactinT^{yr421} receptors that cause the formation of invadopodia and secretion of matrix metalloproteinases [90]. Hence, ASPN-mediated signaling is critical in colorectal cancer metastasis.

3.3.2.3 CD44 Signaling Pathways

CD44 is a non-kinase transmembrane glycoprotein. It is known to induce gastrointestinal, colorectal, prostate, pancreatic, and breast cancer (*as described in* Chap. 2). It is also known to induce head and neck squamous cell carcinoma [91, 92]. CD44 exerts activation of pathways involved in tumor progression via interactions with several ligands, including matrix metalloproteinases (MMPs), osteopontin

(OPN), and hyaluronic acid (HA) [92]. ASPN is an important ligand of CD44. It enhances the co-invasion of CAFs and cancer cells both in vivo and in vitro through the CD44/Rac1 mediated axis [89]. In pancreatic cancer, ASPN promotes cancer cell metastasis by interacting directly with CD44, and the ASPN-CD44 interaction is known to promote CD44/ERK/NF-κB/p65 and CD44/AKT/NF-κB/p65 axes, which activate EMT in a paracrine or autocrine fashion. This results in the upregulation of vimentin N-catenin, ZEB1, snail and slug while expression of E-cadherin and ZO1 is downregulated [89]. Therefore, ASPN-CD44 signaling pathway can be considered as a potential therapeutic target to decrease tumor progression.

3.3.3 Role of Asporin in Joint Diseases

Osteoarthritis is prevalent among majority of geriatric population across the world. Chronic joint diseases and deteriorated musculoskeletal conditions are the leading cause of morbidity in majority of the elderly population. Recent studies indicate strong connection of ASPN with osteoarthritis possibly involving the TGF-β/ECM complex [70] and overexpression of ASPN is known to facilitate osteoarthritis progression. Thus, inhibition of ASPN by miRNA-26b-5p downregulates osteoarthritis in cartilage by TGF-1/Smads pathway [4]. Further, TGF-β induces ASPN expression through Smad-3 indirectly [93]. Therefore, ASPN is suggested as a potential biomarker associated with osteoarthritis, and other pathologies such as intervertebral disc degeneration linked to dysregulation of mi-RNAs, which suggest that ASPN may be a risk factor for lumbar disc degeneration. It has also been noticed that IL-1 upregulates ASPN expression via the NF-kappaB p65 pathway that contributes to intervertebral disc degeneration [4, 94]. Thus, ASPN modulate TGF-β/Smads pathway to inhibit the deposition of collagen in bone and joints, and these changes ultimately precipitates as joint tissue degeneration or osteoarthritis.

3.3.4 Role of Asporin in Cardiovascular Diseases (CVDs)

Structural and functional changes in myocardium and vascular system with age is inevitable [95]. Epithelial-mesenchymal transition (EMT) is primarily responsible for occurrence of CVDs, including myocardial fibrosis, myocardial infarction and cardiac or vascular remodeling [96]. EMT mechanisms are associated with ASPN overexpression, and especially in cardiac fibrosis, it is induced by TGF-β, where upregulation of ASPN is linked with glycated low-density lipoproteins, cardiomyocyte apoptosis, hypertrophy, and fibrosis [4]. Further, EMT-related signalling pathways are important in embryonic heart development while cardiac myocyte repairs are dependent on epicardial EMT. On the other hand, LINC00636, an antifibrotic molecule found in human pericardial fluid, suppresses MAPK1 via miR-450a-2-3p overexpression and improves cardiac fibrosis in patients with atrial fibrillation [97]. Hence, ASPN is involved in the process of epithelial-mesenchymal transformation and induces CVDs via pathways as indicated above.

3.3.5 Role of Asporin in Keloid Disorder

Keloid is a fibroproliferative skin disorder characterized by the abnormal accumulation of collagen in the extracellular matrix (ECM), leading to excessive scar tissue formation. Unlike normal scars, keloids extend beyond the original wound boundaries and do not regress over time. This condition involves dysregulated wound healing, where fibroblasts become hyperactive and produce excessive ECM components, particularly type I and III collagen [98]. Keloid progression into cutaneous tumors is linked to TGF-β signalling, and many of the current research points out that ASPN is the most strongly expressed gene in keloids [99, 100], as the conversion of dermal fibroblasts into α-SMA (smooth muscle actin) myofibroblasts is almost negligible in keloids [101]. ASPN- overexpressing human dermal cells in murine wounds show keloid collagen disorders as a histopathological characteristic [77]. Hence, it can be inferred that ASPN inhibits collagen matrix-mediated intercellular communications in fibroblasts during keloid progression.

3.4 Conclusion

Small leucine-rich proteoglycans play a diverse role in modulating various cellular processes. In depth research on molecular mechanisms and elaborate pathways has paved the way into its therapeutic and translational potential. Therefore, continued research into various pathophysiological aspects of SLRPs will undeniably expand the repertoire of therapeutic options and tremendously improve prognosis across a range of medical conditions.

References

1. Henry SP, Takanosu M, Boyd TC, Mayne PM, Eberspaecher H, Zhou W, De Crombrugghe B, Höök M, Mayne R. Expression pattern and gene characterization of asporin: a newly discovered member of the leucine-rich repeat protein family. J Biol Chem. 2001;276(15):12212–21.
2. Nikitovic D, Aggelidakis J, Young MF, Iozzo RV, Karamanos NK, Tzanakakis GN. The biology of small leucine-rich proteoglycans in bone pathophysiology. J Biol Chem. 2012;287(41):33926–33.
3. Bianco PA, Fisher LW, Young MF, Termine JD, Robey PG. Expression and localization of the two small proteoglycans biglycan and decorin in developing human skeletal and non-skeletal tissues. J Histochem Cytochem. 1990;38(11):1549–63.
4. Fan R, Yan X, Zhang W. Relationship between asporin and extracellular matrix behavior: a literature review. Medicine. 2022;101(51):e32490.
5. Schaefer L, Iozzo RV. Biological functions of the small leucine-rich proteoglycans: from genetics to signal transduction. J Biol Chem. 2008;283(31):21305–9.
6. Chen S, Birk DE. The regulatory roles of small leucine-rich proteoglycans in extracellular matrix assembly. FEBS J. 2013;280(10):2120–37.
7. Zeng-Brouwers J, Pandey S, Trebicka J, Wygrecka M, Schaefer L. Communications via the small leucine-rich proteoglycans: molecular specificity in inflammation and autoimmune diseases. J Histochem Cytochem. 2020;68(12):887–906.

8. Merline R, Schaefer RM, Schaefer L. The matricellular functions of small leucine-rich proteoglycans (SLRPs). J Cell Commun Signal. 2009;3:323–35.
9. Ni GX, Li Z, Zhou YZ. The role of small leucine-rich proteoglycans in osteoarthritis pathogenesis. Osteoarthr Cartil. 2014;22(7):896–903.
10. Hua R, Jiang JX. Small leucine-rich proteoglycans in physiological and biomechanical function of bone. Matrix Biol Plus. 2021;11:100063.
11. Nastase MV, Young MF, Schaefer L. Biglycan: a multivalent proteoglycan providing structure and signals. J Histochem Cytochem. 2012;60(12):963–75.
12. Xu T, Bianco P, Fisher LW, Longenecker G, Smith E, Goldstein S, Bonadio J, Boskey A, Heegaard AM, Sommer B, Satomura K. Targeted disruption of the biglycan gene leads to an osteoporosis-like phenotype in mice. Nat Genet. 1998;20(1):78–82.
13. Gesteira TF, Verma S, Coulson-Thomas VJ. Small leucine rich proteoglycans: biology, function and their therapeutic potential in the ocular surface. Ocul Surf. 2023;29:521–36.
14. Lechner BE, Lim JH, Mercado ML, Fallon JR. Developmental regulation of biglycan expression in muscle and tendon. Muscle Nerve. 2006;34(3):347–55.
15. Liu Y, Li W, Li X, Tai Y, Lü Q, Yang N, Jiang J. Expression and significance of biglycan in endometrial cancer. Arch Gynecol Obstet. 2014;289:649–55.
16. Zhu YH, Yang F, Zhang SS, Zeng TT, Xie X, Guan XY. High expression of biglycan is associated with poor prognosis in patients with esophageal squamous cell carcinoma. Int J Clin Exp Pathol. 2013;6(11):2497.
17. Theocharis AD, Skandalis SS, Neill T, Multhaupt HA, Hubo M, Frey H, Gopal S, Gomes A, Afratis N, Lim HC, Couchman JR, Filmus J, Sanderson RD, Schaefer L, Iozzo RV, Karamanos NK. Insights into the key roles of proteoglycans in breast cancer biology and translational medicine. Biochim Biophys Acta Rev Cancer. 2015;1855(2):276–300.
18. Kram V, Shainer R, Jani P, Meester JA, Loeys B, Young MF. Biglycan in the skeleton. J Histochem Cytochem. 2020;68(11):747–62.
19. Bi Y, Nielsen KL, Kilts TM, Yoon A, Karsdal MA, Wimer HF, Greenfield EM, Heegaard AM, Young MF. Biglycan deficiency increases osteoclast differentiation and activity due to defective osteoblasts. Bone. 2006;38(6):778–86.
20. Young MF, Bi Y, Ameye L, Xu T, Wadhwa S, Heegaard A, Kilts T, Chen X. Small leucine-rich proteoglycans in the aging skeleton. J Musculoskelet Nueronal Interact. 2006;6(4):364.
21. Hua R, Ni Q, Eliason TD, Han Y, Gu S, Nicolella DP, Wang X, Jiang JX. Biglycan and chondroitin sulfate play pivotal roles in bone toughness via retaining bound water in bone mineral matrix. Matrix Biol. 2020;94:95–109.
22. Berendsen AD, Pinnow EL, Maeda A, Brown AC, McCartney-Francis N, Kram V, Owens RT, Robey PG, Holmbeck K, de Castro LF, Kilts TM. Biglycan modulates angiogenesis and bone formation during fracture healing. Matrix Biol. 2014;35:223–31.
23. Zanotti S, Negri T, Cappelletti C, Bernasconi P, Canioni E, Di Blasi C, Pegoraro E, Angelini C, Ciscato P, Prelle A, Mantegazza R, Mora M. Decorin and biglycan expression is differentially altered in several muscular dystrophies. Brain. 2005;128(11):2546–55.
24. Brandan E, Cabello-Verrugio C, Vial C. Novel regulatory mechanisms for the proteoglycans decorin and biglycan during muscle formation and muscular dystrophy. Matrix Biol. 2008;27(8):700–8.
25. Amenta AR, Yilmaz A, Bogdanovich S, McKechnie BA, Abedi M, Khurana TS, Fallon JR. Biglycan recruits utrophin to the sarcolemma and counters dystrophic pathology in mdx mice. Proc Natl Acad Sci. 2011;108(2):762–7.
26. Brandan E, Gutierrez J. Role of proteoglycans in the regulation of the skeletal muscle fibrotic response. FEBS J. 2013;280(17):4109–17.
27. Aranda-Rivera AK, Srivastava A, Cruz-Gregorio A, Pedraza-Chaverri J, Mulay SR, Scholze A. Involvement of inflammasome components in kidney disease. Antioxidants. 2022;11(2):246.
28. Gubbiotti MA, Vallet SD, Ricard-Blum S, Iozzo RV. Decorin interacting network: a comprehensive analysis of decorin-binding partners and their versatile functions. Matrix Biol. 2016;55:7–21.

29. Mohan R, Ck Tovey J, Gupta R, Sharma A, Tandon AJ. Decorin biology, expression, function and therapy in the cornea. Curr Mol Med. 2011;11(2):110–28.
30. Dong Y, Zhong J, Dong L. The role of decorin in autoimmune and inflammatory diseases. J Immunol Res. 2022;2022(1):1283383.
31. Iozzo RV. Matrix proteoglycans: from molecular design to cellular function. Annu Rev Biochem. 1998;67(1):609–52.
32. Järveläinen H, Sainio A, Wight TN. Pivotal role for decorin in angiogenesis. Matrix Biol. 2015;43:15–26.
33. Feugaing DD, Götte M, Viola M. More than matrix: the multifaceted role of decorin in cancer. Eur J Cell Biol. 2013;92(1):1–1.
34. Gubbiotti MA, Neill T, Frey H, Schaefer L, Iozzo RV. Decorin is an autophagy-inducible proteoglycan and is required for proper in vivo autophagy. Matrix Biol. 2015;48:14–25.
35. Moscatello DK, Santra M, Mann DM, McQuillan DJ, Wong AJ, Iozzo RV. Decorin suppresses tumor cell growth by activating the epidermal growth factor receptor. J Clin Invest. 1998;101(2):406–12.
36. Neill T, Painter H, Buraschi S, Owens RT, Lisanti MP, Schaefer L, Iozzo RV. Decorin antagonizes the angiogenic network: concurrent inhibition of Met, hypoxia inducible factor 1α, vascular endothelial growth factor A, and induction of thrombospondin-1 and TIMP3. J Biol Chem. 2012;287(8):5492–506.
37. Järveläinen H, Puolakkainen P, Pakkanen S, Brown EL, Höök M, Iozzo RV, Sage EH, Wight TN. A role for decorin in cutaneous wound healing and angiogenesis. Wound Repair Regen. 2006;14(4):443–52.
38. Rühland C, Schönherr E, Robenek H, Hansen U, Iozzo RV, Bruckner P, Seidler DG. The glycosaminoglycan chain of decorin plays an important role in collagen fibril formation at the early stages of fibrillogenesis. FEBS J. 2007;274(16):4246–55.
39. Goetsch KP, Niesler CU. The extracellular matrix regulates the effect of decorin and transforming growth factor beta-2 (TGF-β2) on myoblast migration. Biochem Biophys Res Commun. 2016;479(2):351–7.
40. Edwards IJ, Xu H, Wright MJ, Wagner WD. Interleukin-1 upregulates decorin production by arterial smooth muscle cells. Arterioscler Thromb. 1994;14(7):1032–9.
41. Baghy K, Reszegi A, Tátrai P, Kovalszky I. Decorin in the tumor microenvironment. In: Tumor microenvironment: extracellular matrix components–Part B. Cham: Springer; 2020. p. 17–38.
42. Kishore V, Paderi JE, Akkus A, Smith KM, Balachandran D, Beaudoin S, Panitch A, Akkus O. Incorporation of a decorin biomimetic enhances the mechanical properties of electrochemically aligned collagen threads. Acta Biomater. 2011;7(6):2428–36.
43. Mauviel A, Santra M, Chen YQ, Uitto J, Iozzo RV. Transcriptional regulation of decorin gene expression: induction by quiescence and repression by tumor necrosis factor-alpha. J Biol Chem. 1995;270(19):11692–700.
44. Kuroda K, Shinkai H. Downregulation of decorin expression in dermal fibroblasts by interleukin-4. Arch Dermatol Res. 1997;289:476–80.
45. Tufvesson E, Westergren-Thorsson G. Alteration of proteoglycan synthesis in human lung fibroblasts induced by interleukin-1β and tumor necrosis factor-α. J Cell Biochem. 2000;77(2):298–309.
46. Frey H, Schroeder N, Manon-Jensen T, Iozzo RV, Schaefer L. Biological interplay between proteoglycans and their innate immune receptors in inflammation. FEBS J. 2013;280(10):2165–79.
47. Marra F. Renaming cytokines: MCP-1, major chemokine in pancreatitis. Gut. 2005;54(12):1679–81.
48. Bocian C, Urbanowitz AK, Owens RT, Iozzo RV, Götte M, Seidler DG. Decorin potentiates interferon-γ activity in a model of allergic inflammation. J Biol Chem. 2013;288(18):12699–711.

49. Jin H, Kumar L, Mathias C, Zurakowski D, Oettgen H, Gorelik L, Geha R. Toll-like receptor 2 is important for the TH1 response to cutaneous sensitization. J Allergy Clin Immunol. 2009;123(4):875–82.
50. Soria-Valles C, Gutiérrez-Fernández A, Guiu M, Mari B, Fueyo A, Gomis RR, López-Otín C. The anti-metastatic activity of collagenase-2 in breast cancer cells is mediated by a signaling pathway involving decorin and miR-21. Oncogene. 2014;33(23):3054–63.
51. Sun H, Wang X, Wang X, Xu M, Sheng W. The role of cancer-associated fibroblasts in tumorigenesis of gastric cancer. Cell Death Dis. 2022;13(10):874.
52. Mao X, Xu J, Wang W, Liang C, Hua J, Liu J, Zhang B, Meng Q, Yu X, Shi S. Crosstalk between cancer-associated fibroblasts and immune cells in the tumor microenvironment: new findings and future perspectives. Mol Cancer. 2021;20:1–30.
53. De Visser KE, Joyce JA. The evolving tumor microenvironment: from cancer initiation to metastatic outgrowth. Cancer Cell. 2023;41(3):374–403.
54. Neill T, Schaefer L, Iozzo RV. Decorin as a multivalent therapeutic agent against cancer. Adv Drug Deliv Rev. 2016;97:174–85.
55. Diehl V, Huber LS, Trebicka J, Wygrecka M, Iozzo RV, Schaefer L. The role of decorin and biglycan signaling in tumorigenesis. Front Oncol. 2021;11:801801.
56. Järvinen TA, Prince S. Decorin: a growth factor antagonist for tumor growth inhibition. Biomed Res Int. 2015;2015(1):654765.
57. Khan GA, Girish GV, Lala N, Di Guglielmo GM, Lala PK. Decorin is a novel VEGFR-2-binding antagonist for the human extravillous trophoblast. Mol Endocrinol. 2011;25(8):1431–43.
58. Buraschi S, Neill T, Iozzo RV. Decorin is a devouring proteoglycan: remodeling of intracellular catabolism via autophagy and mitophagy. Matrix Biol. 2019;75:260–70.
59. Järvinen TA, Ruoslahti E. Targeted antiscarring therapy for tissue injuries. Adv Wound Care. 2013;2(2):50–4.
60. Zhang W, Ge Y, Cheng Q, Zhang Q, Fang L, Zheng J. Decorin is a pivotal effector in the extracellular matrix and tumour microenvironment. Oncotarget. 2018;9(4):5480.
61. Reese SP, Underwood CJ, Weiss JA. Effects of decorin proteoglycan on fibrillogenesis, ultrastructure, and mechanics of type I collagen gels. Matrix Biol. 2013;32(7–8):414–23.
62. Goyal A, Neill T, Owens RT, Schaefer L, Iozzo RV. Decorin activates AMPK, an energy sensor kinase, to induce autophagy in endothelial cells. Matrix Biol. 2014;34:46–54.
63. Gubbiotti MA, Iozzo RV. Proteoglycans regulate autophagy via outside-in signaling: an emerging new concept. Matrix Biol. 2015;48:6.
64. Pang X, Dong N, Zheng Z. Small leucine-rich proteoglycans in skin wound healing. Front Pharmacol. 2020;10:1649.
65. Järvinen TA, Ruoslahti E. Target-seeking antifibrotic compound enhances wound healing and suppresses scar formation in mice. Proc Natl Acad Sci. 2010;107(50):21671–6.
66. Lorenzo P, Aspberg A, Önnerfjord P, Bayliss MT, Neame PJ, Heinegård D. Identification and characterization of asporin: a novel member of the leucine-rich repeat protein family closely related to decorin and biglycan. J Biol Chem. 2001;276(15):12201–11.
67. McEwan PA, Scott PG, Bishop PN, Bella J. Structural correlations in the family of small leucine-rich repeat proteins and proteoglycans. J Struct Biol. 2006;155(2):294–305.
68. Iozzo RV, Schaefer L. Proteoglycan form and function: a comprehensive nomenclature of proteoglycans. Matrix Biol. 2015;42:11–55.
69. Sobhan MR, Mehdinejad M, Jamaladini MH, Mazaheri M, Zare-Shehneh M, Neamatzadeh H. Association between aspartic acid repeat polymorphism of the asporin gene and risk of knee osteoarthritis: a systematic review and meta-analysis. Acta Orthop Traumatol Turc. 2017;51(5):409–15.
70. Ikegawa S. Expression, regulation and function of asporin, a susceptibility gene in common bone and joint diseases. Curr Med Chem. 2008;15(7):724–8.
71. Zhan S, Li J, Ge W. Multifaceted roles of asporin in cancer: current understanding. Front Oncol. 2019;9:948.

72. Gruber HE, Ingram JA, Hoelscher GL, Zinchenko N, Hanley EN, Sun Y. Asporin, a susceptibility gene in osteoarthritis, is expressed at higher levels in the more degenerate human intervertebral disc. Arthritis Res Ther. 2009;11:1–7.
73. Yamada S, Tomoeda M, Ozawa Y, Yoneda S, Terashima Y, Ikezawa K, Ikegawa S, Saito M, Toyosawa S, Murakami S. PLAP-1/asporin, a novel negative regulator of periodontal ligament mineralization. J Biol Chem. 2007;282(32):23070–80.
74. Huang C, Sharma A, Thakur R, Rai D, Katiki M, de Freitas Germano J, Song Y, Singh S, Sin J, Sengstock D, Andres AM. Asporin, an extracellular matrix protein, is a beneficial regulator of cardiac remodeling. Matrix Biol. 2022;110:40–59.
75. Huang S, Lai X, Yang L, Ye F, Huang C, Qiu Y, Lin S, Pu L, Wang Z, Huang W. Asporin promotes TGF-β–induced lung myofibroblast differentiation by facilitating Rab11-dependent recycling of TβRI. Am J Respir Cell Mol Biol. 2022;66(2):158–70.
76. Xu L, Li Z, Liu SY, Xu SY, Ni GX. Asporin and osteoarthritis. Osteoarthr Cartil. 2015;23(6):933–9.
77. Liu L, Yu H, Long Y, You Z, Ogawa R, Du Y, Huang C. Asporin inhibits collagen matrix-mediated intercellular mechanocommunications between fibroblasts during keloid progression. FASEB J. 2021;35(7):e21705.
78. Maris P, Blomme A, Palacios AP, Costanza B, Bellahcène A, Bianchi E, Gofflot S, Drion P, Trombino GE, Di Valentin E, Cusumano PG. Asporin is a fibroblast-derived TGF-β1 inhibitor and a tumor suppressor associated with good prognosis in breast cancer. PLoS Med. 2015;12(9):e1001871.
79. Li H, Zhang Z, Chen L, Sun X, Zhao Y, Guo Q, Zhu S, Li P, Min L, Zhang S. Cytoplasmic Asporin promotes cell migration by regulating TGF-β/Smad2/3 pathway and indicates a poor prognosis in colorectal cancer. Cell Death Dis. 2019;10(2):109.
80. Wang Q, Xiong F, Wu G, Wang D, Liu W, Chen J, Qi Y, Wang B, Chen Y. SMAD proteins in TGF-β signalling pathway in cancer: regulatory mechanisms and clinical applications. Diagnostics. 2023;13(17):2769.
81. Anderson NM, Simon MC. The tumor microenvironment. Curr Biol. 2020;30(16):R921–5.
82. Angioni R, Sánchez-Rodríguez R, Viola A, Molon B. TGF-β in cancer: metabolic driver of the Tolerogenic crosstalk in the tumor microenvironment. Cancers. 2021;13(401):2021.
83. Zarzynska JM. Two faces of TGF-beta1 in breast cancer. Mediat Inflamm. 2014;2014(1):141747.
84. Kou I, Nakajima M, Ikegawa S. Binding characteristics of the osteoarthritis-associated protein asporin. J Bone Miner Metab. 2010;28:395–402.
85. Asati V, Mahapatra DK, Bharti SK. PI3K/Akt/mTOR and Ras/Raf/MEK/ERK signaling pathways inhibitors as anticancer agents: structural and pharmacological perspectives. Eur J Med Chem. 2016;109:314–41.
86. Scaltriti M, Baselga J. The epidermal growth factor receptor pathway: a model for targeted therapy. Clin Cancer Res. 2006;12(18):5268–72.
87. Wee P, Wang Z. Epidermal growth factor receptor cell proliferation signaling pathways. Cancers. 2017;9(5):52.
88. Sigismund S, Avanzato D, Lanzetti L. Emerging functions of the EGFR in cancer. Mol Oncol. 2018;12(1):3–20.
89. Zhang Z, Li H, Zhao Y, Guo Q, Yu Y, Zhu S, Zhang S, Min L, Li P. Asporin promotes cell proliferation via interacting with PSMD2 in gastric cancer. Front Biosci Landmark. 2019;24(6):1178–89.
90. Wu H, Jing X, Cheng X, He Y, Hu L, Wu H, Ye F, Zhao R. Asporin enhances colorectal cancer metastasis through activating the EGFR/src/cortactin signaling pathway. Oncotarget. 2016;7(45):73402.
91. Chen C, Zhao S, Karnad A, Freeman JW. The biology and role of CD44 in cancer progression: therapeutic implications. J Hematol Oncol. 2018;11:1–23.
92. Hassn Mesrati M, Syafruddin SE, Mohtar MA, Syahir A. CD44: a multifunctional mediator of cancer progression. Biomol Ther. 2021;11(12):1850.

93. Kou I, Nakajima M, Ikegawa S. Expression and regulation of the osteoarthritis-associated protein asporin. J Biol Chem. 2007;282(44):32193–9.
94. Wang S, Liu C, Sun Z, Yan P, Liang H, Huang K, Li C, Tian J. IL-1β increases asporin expression via the NF-κB p65 pathway in nucleus pulposus cells during intervertebral disc degeneration. Sci Rep. 2017;7(1):4112.
95. Safar M. Ageing and its effects on the cardiovascular system. Drugs. 1990;39(Suppl 1):1–8.
96. Anbara T, Sharifi M, Aboutaleb N. Endothelial to mesenchymal transition in the cardiogenesis and cardiovascular diseases. Curr Cardiol Rev. 2020;16(4):306–14.
97. Zhang K, Wu M, Qin X, Wen P, Wu Y, Zhuang J. Asporin is a potential promising biomarker for common heart failure. DNA Cell Biol. 2021;40(2):303–15.
98. Limandjaja GC, Belien JM, Scheper RJ, Niessen FB, Gibbs S. Hypertrophic and keloid scars fail to progress from the CD34−/α-smooth muscle actin (α-SMA)+ immature scar phenotype and show gradient differences in α-SMA and p16 expression. Br J Dermatol. 2020;182(4):974–86.
99. Macarak EJ, Wermuth PJ, Rosenbloom J, Uitto J. Keloid disorder: fibroblast differentiation and gene expression profile in fibrotic skin diseases. Exp Dermatol. 2021;30(1):132–45.
100. Hawash AA, Ingrasci G, Nouri K, Yosipovitch G. Pruritus in keloid scars: mechanisms and treatments. Acta Derm Venereol. 2021;101(10):adv00582.
101. Limandjaja GC, Niessen FB, Scheper RJ, Gibbs S. The keloid disorder: heterogeneity, histopathology, mechanisms and models. Front Cell Dev Biol. 2020;8:360.

Regulatory Role and Significance of Class II SLRPs: Fibromodulin, Lumican, Osteoadherin, PRELP, and Keratocan

4

Polly Lama and Binod Kumar Tamang

Abstract

This chapter describes the proposition of class II SLRPs fibromodulin, lumican, osteoadherin, proline/arginine leucine-rich repeat protein (PRELP), and keratocan in matrix homeostasis and catabolism. So far, there are only a few published literature that collectively analyses the mechanisms, class II SLRP molecules undergo during various types of pathological changes occurring in the matrix including their potential to act as damage-associated molecular pattern recognition molecules (DAMPs); however, there are also many independent studies assessing the significance of lumican and fibromodulin in tumor formation and in diseases of cartilage, bone, muscle, and skin indicating their potential to act as biomarkers of disease. Thus, this chapter explores the spatiotemporal expression analysis of class II SLRPs and provides significant clues toward understanding the regulatory roles these molecules play in development, diseases, as well as in the maintenance of tissue structure in adult life, thus this chapter aims to facilitate an in-depth understanding on the importance of class II SLRP molecules in the extracellular matrix.

Keywords

Fibromodulin · Lumican · Regulation · Arthritis · Immunity · Cells

P. Lama (✉) · B. K. Tamang
Department of Anatomy, Sikkim Manipal Institute of Medical Sciences, Sikkim Manipal University, Tadong, India
e-mail: polly.lama@smims.smu.edu.in

P. Lama (ed.), *The Biological Role of Small Leucine-Rich Proteoglycans (SLRPs) in Matrix Homeostasis and Diseases*,
https://doi.org/10.1007/978-981-96-7890-7_4

4.1 Expression of Class II SLRPs

Expression of class II SLRPs is dynamic, and differences in spatiotemporal patterns during development, adult life, and various pathological diseases indicate functional changes [1]. Class II SLRPs are usually secreted in the extracellular matrix, and they can interact with other matrix proteins and cell receptors. SLRPs in general are categorized as a major structural component of the matrix modulating synthesis, assembly, and degradation of collagen fibrils, and are also recognized as regulators of cell-matrix interaction, influencing variety of biological processes, such as cell proliferation, differentiation, survival, adhesion, inflammation, angiogenesis, and tumorigenesis. Most of the SLRPs that modulate collagen fibrillogenesis belong to class I and class II group [2, 3].

4.2 Fibromodulin

Fibromodulin is a keratan sulfate-binding SLRP molecule that interacts with collagen types I, II, and regulates fibrillogenesis, modulating micro-mechanical effects in the peripheral cornea, sclera, skin, and tendon [4]. Fibromodulin-deficent animal model shows an abnormal connective tissue cell phenotype with impaired formation of collagen fibrils [5, 6]. The molecular structure of fibromodulin has several aspartic and glutamic acid residues with a variable number of tyrosine clusters attached to the N-terminus that are O-sulfated, and these residue side chain changes during articular cartilage growth and development [7, 8]. Fibromodulin is detected in various types of connective tissues, and they are one of the most potent collagen-binding SLRPs in cartilage, where its presence correlates with the size of collagen fibrils. The temporospatial transcriptional expression pattern shows it exists with type II collagen and are also linked to cartilage development and maintenance particularly in the early stage of articular cartilage genesis. Fibromodulin is therefore categorized as a chondrogenic marker, and the fragments of the protein core molecule are a feature of cartilage osteoarthritis, including disc degeneration and temporomandibular joint degeneration, where the fibromodulin core protein degradation predominately occurs at the C-terminus [9], Fig. 4.1.

4.2.1 Regulatory Role of Fibromodulin in Cartilage

Regulatory role of fibromodulin in cartilage especially during osteoarthritis progression has been shown to induce three main biological changes. The *first* mechanism involves direct binding of fibromodulin to complement system and the inhibitor factor H [10]. Intact FMOD blocks complement system from activation but the fragmented molecules that appears during degeneration, injury, arthritis, or infection binds with complement factor C1q at the N-terminus to activate the innate inflammatory pathway essential for elimination of pathogens and damaged cells during tissue recovery and reconstruction [11]. The *second* alternate proposed mechanism

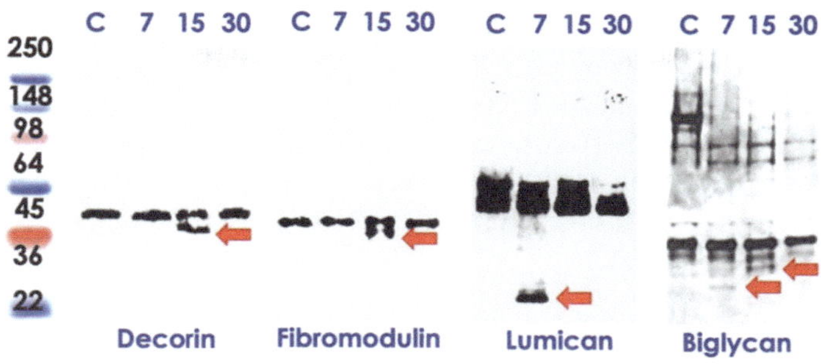

Fig. 4.1 Western blot analysis of decorin, fibromodulin, lumican, and biglycan core protein and their fragmented peptides on a 10% gel. Lane was loaded as "control C, non-degenerated rat disc" first followed by 7, 15, and 30 days loaded rat discs. Decorin core protein had a molecular weight of 45 KDa and a fragmented peptide at 36 KDa, as indicated by the red arrow. Fibromodulin core protein had a molecular weight of 40 KDa followed by a fragment in 15 days loaded rat disc, and this fragment was lost in 30 days loaded rat discs. Lumican core protein band showed at 45 KDa, followed by fragmented 22 KDa peptide in 7 days loaded rat disc. Biglycan appeared as a doublet band with a core protein of 40 KDa, and multiple fragments in 7,15, and 30 days loaded rat disc, which were weakly expressed, indicating its loss from the cartilage tissue. (Figure adapted from Dr. Polly Lama, Molecular Matrix Lab)

prevents polymorphonuclear neutrophils and fibroblasts from adhering to the cartilage surfaces due to the ability of fibromodulin to directly bind to collagen type II, thus protecting cartilage collagen from being recognized by the inflammatory cells and collagenases as a degraded entity [12]. In contrast, accumulation of the cleaved molecules destabilizes the collagen fibrils and exposes the neo-epitopes (new binding sites) to cell receptors and other proetolytic enzymes in arthritis. The *third* mechanism show fibromodulin alter signal transduction events by forming complexes with transforming growth factor beta (TGF-β) superfamily. In matrix homeostasis, fibromodulin protects cartilage tissue from degradation effects initiated by TGF-β/bone morphogenetic protein-2 (BMP-2), and overexpression during matrix imbalance may mitigate arthritis-associated chondrocyte apoptosis [8, 13]. These regulatory factors suggest fibromodulin is a potential therapeutic candidate for osteoarthritis treatment and presents a prospect to optimize its use in clinical application [14].

4.2.2 Regulatory Role of Fibromodulin in Bone

Regulatory role of fibromodulin in bone is demonstrated by its strong affinity to bind with the hydroxyapatite an essential mineral that plays a vital role in maintaining bone strength and regeneration [15]. Fibromodulin is expressed during fetal endochondral and intramembranous bone ossification, induced by bone morphogenetic protein-2 (BMP-2) and which is secreted by the bone-forming osteoblasts cells [9]. Fibromodulin significantly hinders the maturation of osteoclasts that are known to remove bone forming osteoblast cells, suggesting an osteoblast-osteoclast crosstalk in maintaining bone tissue matrix homeostasis [16]; further, decrease in fibromodulin expression is indicated in the aging bones as an inevitable part of osteoporosis, showing the importance of fibromodulin in bone development and maintenance [17]. Analytical studies on the fibromodulin knockout mice model also reveal that it becomes overexpressed in mineralizing trabeculae of the mandible bone and can even result in significant ectopic ossification of the tendons [18]; fibromodulin thus represses bone formation when there is an alteration in the matrix microstructure.

4.2.3 Regulatory Role of Fibromodulin in Dentine Tissue

Regulatory role of fibromodulin in dentine tissue shows its peculiar localization in the periodontium, supra basal gingival epithelium, periodontal ligament, alveolar bone, gingival, and periodontal fibroblast cells, as well as their corresponding matrix [19, 20]. Fibromodulin levels increase during periodontal tissue mineralization, and the knockout mice model of fibromodulin shows abnormal collagen fibril bundles in hypomineralized mandibles and periodontal tissue ligament [21]. Further, fibromodulin and biglycan double-deficient mice exhibit altered remodeling of alveolar bone with increased tartrate-resistant acid phosphatase-positive cells localized in the periodontal ligament [22]. Fibromodulin in predentin tissue restricts collagen fibrillogenesis to promote tissue mineralization and enamel formation [5]. Therefore, fibromodulin knockout mice exhibit delayed enamel formation with twisted and porous tissue appearances. In human dentine tissue pathogenesis, fibromodulin is strongly expressed around carious lesions but is diminished within severely degraded tissue regions [23]. Thus, fibromodulin-based potential treatment target offers a prospect for further investigations.

4.2.4 Regulatory Role of Fibromodulin in Tendon

Regulatory role of fibromodulin in tendon indicates significance during the process of tissue repair mediated by the tendon stem/progenitor cells [24] that are localized between the collagen fibrils. These stem cells express fibromodulin allowing self-renewal of cells, protection from calcium-induced excessive accumulation of BMPs, and ectopic tendon ossification [25]. Fibromodulin recruits tenocytes (cells of the tendon) by stimulating the synthesis of various matrix metalloproteinases (MMPs) that upregulate lysyl oxidase, essential for cross-linking the collagen fibers, imparting tensile strength in tendons [26], and also for promoting healing during tendon injury. Thus, a minimally invasive procedure to deliver fibromodulin in hyaluronic acid granular hydrogel to treat injured tendons has been developed, and the FMOD releasing hydrogel significantly promotes tendon healing in a fully ruptured rat's Achilles tendon. Temporal tendinitis of the temporomandibular joint (TMJ) is the cause of chronic orofacial pain that is often difficult to treat due to the lack of effective cure strategies [27], and the novel treatment method with fibromodulin releasing hydrogel may act as an alternate treatment to combat TMJ disorders.

4.2.5 Regulatory Role of Fibromodulin in Muscles

Regulatory role of fibromodulin in muscles shows a high expression level at the junctions between the tendons and cardiac muscle [28] especially at the fibrous annulus ring, where the semilunar valves anchor the aorta, permitting mesenchymal cell differentiation [29]. Fibromodulin reduces the expression of fibrosis-associated molecules, LOX, in cultured cardiac fibroblast cells that exhibit anti-scarring function during cardiac remodeling [30]. In skeletal muscle, fibromodulin promotes tissue regeneration at the site of injury by activating satellite cells that suppress myostatin produced by myocytes and promote myoblast differentiation; this process allows regulation of calcium channels, allowing calcium influx that stimulates myoblast-to-myotube differentiation [31]. Furthermore, fibromodulin regulates the differentiation of myoblast cells to adipocyte-like cells and prevents lipid accumulation, therefore supporting the myogenesis process and presenting a prospect for developing newer therapeutics to treat different muscular diseases [32].

4.2.6 Regulatory Role of Fibromodulin in Skin

Regulatory role of fibromodulin in skin is particularly significant in cutaneous wound healing process as fibromodulin is the only SLRP that gets downregulated to promote fetal-type scarless healing as opposed to adult-type skin wound healing leading to scar formation [33]. In early gestational stages, fetal rodents show a high level of fibromodulin, which is sufficient for promoting fetal wounds in the late gestational stage. Thus, adult fibromodulin knockout mice demonstrates delayed wound healing, accompanied by prolonged inflammation, impaired angiogenesis, and unorganized collagen architecture, with increased scar formation [34]. Evidence from in vivo and in vitro studies indicate fibromodulin enhances angiogenesis and accelerates endothelial cell adhesion, spreading, migration, invasion, actin fiber formation, including assembly of tubelike networks of vessels, leading to the activation of quiescent endothelial cells [35]. In the early stage of wound healing, fibromodulin also promotes fibroblast migration into the injured sites as myofibroblasts and generates a contractile force around the wound edges against mechanical stimulus, leading to faster wound closure by selectively promoting myofibroblast clearance to prevent excessive deposition and accumulation of contracted collagen fibers [36]. Application of fibromodulin is shown to successfully reduces scar tissue formation without impacting the tensile tissue strength in rodent and porcine models [37]. Thus, the antifibrotic effects of fibromodulin may be applicable to the lung, liver, and other diseased tissues as myofibroblast is the predominant cause of fibrosis involving the matrix in these tissues.

4.2.7 Regulatory Role of Fibromodulin in Cancer

Regulatory role of fibromodulin in cancer involves activation of pro-angiogenesis factor (*as described in* Chap. 2) and this is particularly well evidenced in small cell lung cancer, prostate cancer, and glioblastoma. In addition, fibromodulin modulates pro-migratory inflammation cell invasion in breast cancer and glioblastoma [38, 39], and fibromodulin level is also elevated in B-cell chronic lymphocytic leukemia [40]. Therefore, fibromodulin is considered a target for treating cancerous diseases with the potential to manage localized tumorous in colon carcinoma. In glioblastoma and prostate cancer, high fibromodulin expression is associated with malignancy as it can be detected in blood, urine, and prostatic secretions, and fibromodulin is therefore suggested to be a potential biomarker for tumor diagnosis as well as prognosis [41].

4.3 Lumican (LUM)

Lumican (LUM) is a keratan sulfate-rich SLRP found within the interstitial tissues and acts as a major component of bone tissues [42]. Lumican shares similar sequence homology (38–45%), with fibromodulin and keratocan with overlapping functional properties, and regulates collagen fibrillogenesis in the cornea essential for maintaining optical clarity [43], and lumican knockout mice display severe disruption in matrix organization leading to loss of corneal opacity. Lumican is expressed in malignant tumors, including esophageal, lung, gastrointestinal, breast, colorectal, and pancreatic cancers [44], and, on the contrary, lumican also display antitumor activity [45] by inhibiting cell proliferation in many cancer cell lines, and are also classified as a potent biomarker of matrix diseases [46]. Further in bone and cartilage tissue, lumican contributes to disease processes, Fig. 4.1, and their fragments are considered to be effective for monitoring prospective therapeutic applications. Many other SLRPs and lumican interact with several cell receptors, growth factors, cytokines, and chemokines with potent roles in various pathological tissues affecting cellular proliferation, migration, and differentiation, thus making them a suitable biomarker in tissue pathogenesis.

4.3.1 Regulatory Role of Lumican in Fibrotic Liver and Kidney Diseases

Regulatory role of lumican in fibrotic liver and kidney diseases tend to control the process associated with inflammation and immune responses [47, 48]. In renal (kidney) tissue diseases fibrotic deposition of collagens I, II, III, V, VI, VII, and XV occurs, with upregulation of collagens I and III in the early stages of renal fibrosis [49]. Similarly, in liver fibrosis and hepatocellular carcinoma, increased accumulation of collagen fibers leads to scar tissue formation due to repetitive tissue damage and inflammation [50]. Lumican activates immune responses in the kidney fibrosis and inflammation as it interacts with toll-like receptors. In diabetic nephropathy, an end-stage renal disease lumican gene is associated with the excessive accumulation of extracellular matrix component and thus considered to be a diagnostic biomarker for disease progression [51]. Lumican therefore act as an innate immune receptor, coreceptors, and adaptor molecules, promoting pro- and anti-inflammatory signaling mechanism, which either controls inflammation or accelerates the process of tissue pathogenesis.

4.3.2 Regulatory Role of Lumican in Cardiovascular Diseases

Regulatory role of lumican in cardiovascular diseases is ascribed as upregulated expression in experimental and clinical heart failure [52], where lumican interacts with beta-1,4-galactosyltransferase 5 (B4GALT5) enzyme that accelerates inflammation and regulates cardiac tissue fibrosis by the glycogen synthase kinase-3 (GSK-3) or β-catenin signaling pathway. Cardiac tissue proteins undergo posttranslational glycosylation, leading to increase in B4GALT5 enzyme along with TGFβ1 that activates cardiac fibroblasts, and knockdown of B4GALT5 enzyme decreases the conversion of cardiac fibroblast to myofibroblasts, reducing collagen deposition, whereas overexpression of B4GALT5 elevates cardiac fibroblast activation by the Akt/GSK-3β/β-catenin pathway, promoting fibrosis in congestive heart failure [53]. In liver cancer, B4GALT5 is highly expressed, catalyzing the biosynthesis of lactosylceramide, an intermediate compound formed in the biosynthesis of sphingolipids and gangliosides [54]. A range of cell biological processes are critically modulated by sphingolipids that includes cell growth, migration, adhesion, apoptosis, senescence, and inflammatory responses in neurodegenerative diseases, metabolic disorders, cancer, cardiovascular diseases, and skin disorders [55].

4.3.3 Regulatory Role of Lumican in Retinal Tissue

Regulatory role of lumican in retinal tissue involves maintaining corneal collagen fibril organization essential for sustaining corneal transparency [56], and in the diseases of retina, lumican levels correlate with pro-inflammatory changes that lead to tissue degeneration. Lumican gene presents a single-nucleotide polymorphism in the 5′-regulatory sequence in Asian patients with high myopia [57], but on the contrary, lumican has a protective effect in myopia in cohorts of Caucasian patients; thus, the prevalence of myopia varies among different ethnicities [58]. Lumican level decreases with age-related macular degeneration and induces neovascularization in diabetic retinopathy [59], one of the leading causes of visual impairment in the world. Lumican accelerates inflammation and wound healing, as it permits recruitment of macrophages at the site of corneal injury and activates neutrophils in lipopolysaccharide-induced keratitis in mice and inflammatory uveitis [51]. Increased serum level of lumican is suggested to act as a pathological biomarker indicating fibrocellular degeneration of the vitreoretinal membrane and idiopathic changes in the epiretinal membrane [60]. Thus, lumican level changes in inherited and acquired eye diseases.

4.3.4 Regulatory Role of Lumican in Neurodegenerative Disorder

Regulatory role of lumican in neurodegenerative disorder is particularly evidenced during the development of human neocortex, leading to the formation of brain infolds, where lumican modulates axonal guidance for the proper formation of neural networks in the brain. Therefore, in situ hybridization analysis reveals lumican is expressed exclusively in the lateral cortex and controls inter-axonal crosstalk in the corticospinal tract, facilitating migration and interconnections of the longitudinal axons in establishing functional neural circuits/pathways [61]. Further, lumican gene expression is specifically downregulated by the endothelial cells regulating the blood-brain barrier when disturbance in fluid flow is encountered, showing the significance of lumican in maintaining shear stress at the endothelial cell–matrix junction promoting barrier function, which changes during trauma compromising the cerebrovascular blood flow or in stroke where the endothelial cell junction associates with significant downregulation in lumican gene expression [62].

4.3.5 Regulatory Role of Lumican in Disorders Associated with Pregnancy and Ovulation

Regulatory role of lumican in disorders associated with pregnancy and ovulation is mediated in the female uterine cervix, as reorganization of the collagen fibers occurs during pregnancy and ovulation [63]. Lumican is downregulated in the placenta during pregnancy, promoting preeclampsia of blood vessels, and differences in the spatial localization of vessels is evident in normal and pathological uterine endometrium and polycystic ovary syndrome [64]. Further, lumican expression and regulation in the murine uterus has been shown to regulate estradiol and progesterone hormones that are associated with matrix remodeling in uterine tissues, including degenerative pathological changes [65]; thus, lumican is suggested to act as a potential biomarker and a therapeutic target in treating infertility of the uterine endometrium [66]. Similarly, in polycystic ovarian diseases, lumican levels are significantly elevated resulting in excessive deposition of various other SLRPs, especially decorin and lumican in the uterine endometrium [67].

4.3.6 Regulatory Role of Lumican in Muscles, Bone, and Cartilage

Regulatory role of lumican in muscles, bone, and cartilage involves upregulated expression during musculoskeletal tissue pathogenesis and is proposed to act as a discriminative biomarker in intervertebral disc degeneration and low back pain [68]. Many other classes of SLRPs, including lumican, reflect pathological changes in the intervertebral disc tissue, such as disc space narrowing, vertebral osteophyte formation, and inflammatory upregulation including Modic changes, endplate avulsion, and increased mechanotransduction, leading to loss of matrix organization and pain [69]. Silencing of the lumican gene triggers tumor necrosis factor (TNF)-α-induced inflammatory response, cell senescence, and cell apoptosis, which is comparable to changes as evidenced in a herniated intervertebral disc [70]. Lumican in disc tissue regulates apoptosis of the nucleus pulposus cells by mitogen-activated protein kinase-5/p38 cell signaling pathway that expresses Fas ligand receptor and thus leads to cell death. In bone tissue, lumican has an osteoprotective function due to its ability to simultaneously increase bone formation and decrease bone resorption [71]. Lumican inhibits osteoclastogenesis by suppressing Akt activity but does not inhibit MAP kinases, such as c-Jun N-terminal kinase, p38, and extracellular signal-regulated kinase (ERK). Further, lumican is overexpressed in tendons present at the joints and is associated with fibrosis involving the TGF-β signaling pathway [72]. Finally, in myogenesis, lumican upregulates phosphorylation of p38 mitogen-activated protein kinase pathway (MAPK), protecting muscle wastage [73].

4.3.7 Regulatory Role of Lumican in Cancer

Regulatory role of lumican in cancer correlates with the development of esophageal cancer [74], lung adenocarcinoma and squamous cell carcinoma [75], breast cancer [76], osteosarcoma [77], colorectal carcinoma [78], pancreatic cancer [79], ovarian cancer [66], bone metastasis arising from breast cancer [80], and gastric cancer [81] (*described in detail in* Chap. 2). Thus, lumican promotes tumorigenesis, but it also inhibits tumor development and is shown to suppress pancreatic cancer cell growth by stimulating the growth of normal pancreatic cells and further inhibiting tumor cell migration. In B16F1 melanoma cell line for lung metastasis, lumican inhibits the migration of these cells by blocking MMP-14 activity [45], and similarly, lumican is reported to inhibit the progression of prostate cancer [82]. The antitumor activity of lumican is specific to lumican peptides and its ability to act as an MMP inhibitor, and these peptides also interact with the integrin α2β1 receptor that hinders the development of new blood vessel required for the nutrition of the tumor cells [83].

4.3.8 Regulatory Role of Lumican in Periodontal Diseases

Regulatory role of lumican in periodontal diseases upregulates the prognosis by affecting the organization of collagen fibril in the gums, promoting inflammation of the gingiva, infection of the cementum, and weakening the anchoring periodontal ligaments at the bone-teeth junction [84].

4.4 Osteoadherin, Also Termed Osteomodulin

Osteoadherin, also termed osteomodulin, was first isolated from bovine diaphyseal bone as described in Chaps. 1 and 2; it is a keratan sulfate glycosaminoglycan side chain bearing SLRP that promotes cell attachment and regulates the matrix during bone formation [85]. Bovine osteoadherin acts by binding osteoblasts cells to the integrin alpha-V and beta-3 receptors, which can be inhibited by a synthetic peptide containing the integrin-binding sequence arg-gly-asp. Human osteoadherin binds to type I collagen fibers and suppresses the rate and extent of collagen fibril formation. Osteoadherin can reduce the diameter of collagen fibrils to induce variability, leading to straight collagen fibrils as opposed to the regular twisted orientation [86]. Osteomodulin is also expressed by human dental pulp stem cells (DPSCs), modulating osteo/odontoblastic differentiation in those cells [87].

4.4.1 Regulatory Role of Osteoadherin in Osteoarthritis

Regulatory role of osteoadherin in osteoarthritis in adults involves the formation of subchondral bone sclerosis, which is a key feature of osteoarthritis and considered to create a secondary impact effect on the bone [88]. Thus, in the spontaneous animal model of osteoarthritis, loss of osteoadherin associates with articular cartilage degradation and may even prevent cartilage from developing lesions. Further, in the zebrafish model, osteoadherin prevents the formation of spontaneous cartilage lesions during aging as osteoblasts and hypertrophic chondrocytes decreases, and so does the synthesis of osteoadherin, indicating its deleterious role in cartilage. Therefore, osteoadherin contributes to osteoarthritis by regulating metabolic function at the bone-cartilage junction [89, 90], and transcriptomic data analysis reveals that osteoadherin does not perform its function during direct gene expression by osteoblasts. Further, bone sialoprotein, a mineral component expressed by mature osteoblasts, osteoclasts, and hypertrophic chondrocytes of the growth plate, is downregulated by osteoadherin and overexpressed by hypertrophic chondrocytes in osteoarthritis [91]. Thus, osteoadherin controls cartilage calcification in osteoarthritis by regulating activity of bone sialoprotein.

4.4.2 Regulatory Role of Osteoadherin in Bone

Regulatory role of osteoadherin in bone is profuse due to its ability to bind with hydroxyapatite involved in bone mineralization [92]. Osteoadherin regulates collagen fibril formation in bones as demonstrated by analyzing the recombinant human osteoadherin produced by the baculovirus expression method that also showed osteoadherin interacts with type I collagen fibers, a major bone matrix protein [86], and osteoadherin interacts with purified immobilized collagen, suppressing fibril formation in a turbidity assay. Further, morphological analysis of collagen in the presence or absence of osteoadherin demonstrates that it reduces the diameter and changes the shape of collagen fibrils, thus indicating its significance in the formation of bone tissue [93].

4.4.3 Regulatory Role of Osteoadherin in Dentine Tissue

Regulatory role of osteoadherin in dentine tissue is seen during osteo/odontoblastic differentiation and mineralization process moderated by the human dental pulp stem cells (hDPSCs). These cells show upregulated mRNA expression of osteoadherin regulates dentinogenesis and dental repair. Past studies exploring these mechanisms indicate that expression of osteoadherin is regulated by cytokines TGF-β1 and BMP2, where TGFβ1 downregulates while BMP-2 upregulates osteoadherin [94]. Thus, osteoadherin may play a role in TGFβ and BMP signaling during osteo/odontogenic differentiation, altering the mineralization process and even suppressing the expression of the osteo/odontoblast-related genes.

4.5 Proline/Arginine Leucine-Rich Repeat Protein (PRELP)

Proline/arginine leucine-rich repeat protein (PRELP), also called prolargin, is the only SLRP with proline and arginine residues at the amino-terminal region bearing the heparin and heparan sulfate glycosaminoglycan side chain [95]. PRELP can bind to perlecan, a heparin sulfate proteoglycan of the basement membrane, and collagen types I and II through its LRR repeats, as described in Chap. 1. Based on its pericellular localization near the basement membranes and corresponding matrix, PRELP is suggested to act as a link molecule anchoring the epithelial cell basement membranes to the underlying connective tissue. Thus, in defects of the basement membranes and cartilage, PRELP is characterized by loss in collagen-binding function [96]. Overexpression of PRELP in transgenic mice changes the skin structure and decreases collagen fiber bundles and the hypodermal fat layer, indicating that PRELP may interact with fibroblasts and adipocytes cell-matrix junction [97]. PRELP also inhibits the formation of the complement membrane attack complex (MAC), limiting pathological activation of complement factor in inflammatory diseases such as rheumatoid arthritis [98]. The heparan-binding domain of PRELP additionally impairs the osteoclastogenesis process in bones by NF-κB inhibition, preventing osteoporosis and breast cancer osteolytic metastases in preclinical in vivo models [99], indicating the therapeutic potential of PRELP in various pathological conditions.

4.5.1 Regulatory Role of PRELP in Skin

Regulatory role of PRELP in skin shows it has low expression level, which is in contrast to other connective tissue, such as the cartilages where PLREP is abundant; thus, PRELP overexpression in transgenic mice generated through the human

PRELP transgene shows altered skin connective tissue phenotype with perturbed organization of collagen fibrils in the dermis of the skin and diminished thickness of the hypodermal fat layer [100]. Overexpression of PRELP in skin compromises collagen fiber bundle content and size in the dermis, and PRELP interacts with the collagen-producing fibroblast cells to downregulate protein synthesis of the collagen molecules that participate in fibril formations; however, PRELP itself does not influence the process of collagen fibrillogenesis.

4.5.2 Regulatory Role of PRELP in Complement Immunity and Rheumatoid Arthritis

Regulatory role of PRELP in complement immunity and rheumatoid arthritis shows it limits pathological activation of chronic inflammation in the endothelial cell basement membranes, damaged region of the cartilage, and rheumatoid arthritis [98]. Complement systems are a part of human innate immunity as they confer defense action against pathogens as well as endogenous matrix danger signals [101]. In rheumatoid arthritis, complement activation is triggered by immune complexes, apoptotic/necrotic cells, and released cartilage components, such as the SLRP fragments, which have the potential to activate both the innate and adaptive immunity by promoting interactions between complement products and the target cells. Complement membrane attack complex (MAC) is a cytolytic effector of innate and adaptive immunity that forms pores in the plasma membrane of pathogens or targeted cells, leading to osmolysis. Nucleated cells resist MAC-mediated cytolysis by the expression of inhibitors that block MAC assembly or by removing these complexes through endocytosis. Formation of MAC in arthritis contributes to the pathogenesis, as $CD59a^{-/-}$ mice model develops more severe arthritis than their wild-type counterparts [102]. PRELP inhibits MAC by decreasing the polymerization of complement protein C9 and by preventing lytic pore formation. PRELP also inhibits the alternative complement stage involving C3-convertase that inhibits properdin and factor B into forming convertase. PRELP has also been shown to bind to C4BP, a specific type of complement factor that binds to necrotic cell DNA, inhibiting complement activation in rheumatoid arthritis, and in degraded cartilage PRELP, acts as a inhibitor by protecting it from further complement deposition [98]. Therefore, PRELP has multiple targets in complement inhibition by directly involving the C3-convertase, MAC expressions, and by recruting C4BP at sites with already exposed PRELP.

4.5.3 Regulatory Role of PRELP in Bone

Regulatory role of PRELP in bone involves inhibition of osteoclastogenesis, a process that breaks down or absorbs bone tissue during skeletal remodeling [99]. As demonstrated by an in vitro mouse model, the N-terminal region of PRLEP that binds with the heparin sulfate side chain [103] is broken down and gets internalized through the chondroitin sulfate and annexin II-dependent mechanism involving nuclear translocation and interaction with the p65 nuclear factor κB (NF-κB) complex [104], inhibiting DNA-binding transcriptional activity leading to downregulation of osteoclast-specific genes. Thus, intact PRELP core protein impairs osteoclastogenesis but does not disrupt the mitogen-activated protein kinase signalling nor does it impair survival. PRELP is also internalized by the RAW264.7 osteoclast-like cell line that fails to affect calvarial osteoblasts, bone marrow macrophages, and epithelial cell lines. In vivo, heparin-binding PRELP reduces osteoclast number and activity in ovariectomized mice, suggesting its physiological and pathological importance in bone remodeling [105].

4.6 Keratocan (KERA)

Keratocan (KERA) are the regulators of corneal stroma in matrix organization and assembly involving collagen fibres, and presents a high sequence similarity to lumican [106]. In adult tissues, keratocan expression is considered a phenotypic marker for keratocytes (a corneal fibroblast cell-type) present in corneal stroma [107]. Lumican is a regulatory protein for keratocan expression at the promoter level in the adult mouse corneal stroma, and the expression level of keratocan decreases along with the lumican, and both of these SLRPs are essential for maintaining matrix equilibrium and tissue transparency in the cornea [56]. Experimental animal model shows lumican-null mice develop cloudy corneas due to an altered collagenous matrix structure characterized by larger fibril diameters and disorganized spacing, and in contrast, keratocan knockout mice exhibit a thin but clear cornea with insignificant alteration of stromal collagenous matrix [106]. Mutations of keratocan cause corneal plana in humans, which is often associated with glaucoma [108]. These findings show that lumican and keratocan have different roles in regulating the formation of stromal extracellular matrix, as many experimental animal models indicate that lumican has additional biological functions, such as regulation of cell migration and epithelium-mesenchyme transition in wound healing and tumorigenesis, besides regulating collagen fibrillogenesis (*described in* Chap. 2).

4.6.1 Regulatory Role of Keratocan in Cornea Plana

Regulatory role of keratocan in cornea plana causes an abnormally flat shape cornea, leading to a decrease in refraction as the normal angle of protrusion between the cornea and the sclera is missing [109]. Reduced corneal curvature can lead to hyperopia, a hazy corneal limbus, and arcus lipoides at an early age [110]. Corneal plana is categorized into two types: a more severe, recessively inherited form as CNA2 and a milder, dominantly inherited form as CNA1. CNA2 is a rare disorder present globally with high prevalence in the Finnish population [111]. The gene mutated in CNA2 is located at 12q [112], where a cluster of several SLRP genes exists [113, 114]. In cornea, the KERA gene encoding keratocan is located at 12q, and mutations in the KERA gene in CNA2 Finnish patients show missense code, leading to the substitution of a highly conserved amino acid keratocan protein sequence, while in American patients, homozygous mutation of CNA2 show premature stop codon that truncates the KERA protein. Thus, a mutation in KERA gene causes CNA2, while patients with CNA1 show no mutations in the SLRP genes [108].

4.6.2 Regulatory Role of Keratocan in Osteogenic Differentiation of Bone Cells

Regulatory role of keratocan in osteogenic differentiation of bone cells involves proliferation and modulation of osteoprogenitor lineage differentiation [115]. Gene expression analyses of bone cells in different stages of maturation indicate that keratocan is differentially expressed by osteoblasts, while the osteocytes show no expression. Calvarial osteoblast cultures derived from keratocan null mice shows a slower mineralization process than in the wild-type osteoblast cultures, and mRNA derived from keratocan knockout mice model demonstrates a reduction in mature osteoblast differentiation markers, bone sialoprotein (BSP) and osteocalcin. Keratocan-deficient homozygous knockout mice have significantly decreased rates of bone formation and mineral deposition [106]. Thus, these findings show that keratocan has the ability to regulate osteoblast function.

4.6.3 Regulatory Role of Keratocan in Articular Cartilage

Regulatory role of keratocan in articular cartilage show a catabolic effect, as the keratocan core protein becomes increasingly fragmented as the cartilage tissue degenerates. Keratocan core proteins have a molecular size of 35–37 kDa that generates four core protein fragments of relatively small size between 14 and 25 kDa in meniscus and fibrillated OA cartilage [116]. Fragments of all SLRPs increase with age, except for keratocan in the knee articular cartilage. Thus, one of the most distinct changes exhibited by keratocan in the knee articular cartilage is the decrease in the size of the intact core protein. Increased SLRP fragments can act as a biomarker to evaluate cartilage tissue metabolism or disease progression in spondylarthritis and osteoarthritis in humans and animal models [117].

SLRP fragments that are retained in the diseased matrix in knee meniscus and hip articular cartilages can be further analyzed to determine the specific cleavage site of the core protein generating these fragments, and this could lead to the identification and development of antibodies that could recognize the specific amino and carboxyl terminus. These antibodies can be further used to discover potential biomarker peptides that can be released into the body fluids.

References

1. Shu CC, Flannery CR, Little CB, Melrose J. Catabolism of Fibromodulin in developmental rudiment and pathologic articular cartilage demonstrates novel roles for MMP-13 and ADAMTS-4 in C-terminal processing of SLRPs. Int J Mol Sci. 2019;20(3):579.
2. Dai G, Sun Y, Wei R, Xi L. Small leucine-rich proteoglycan. Curr Issues Mol Biol. 2023;45(7):6116–39.
3. McEwan PA, Scott PG, Bishop PN, Bella J. Structural correlations in the family of small leucine-rich repeat proteins and proteoglycans. J Struct Biol. 2006;155(2):294–305.
4. Chen S, Oldberg A, Chakravarti S, Birk DE. Fibromodulin regulates collagen fibrillogenesis during peripheral corneal development. Dev Dyn. 2010;239(3):844–54.
5. Goldberg M, Septier D, Oldberg A, Young MF, Ameye LG. Fibromodulin-deficient mice display impaired collagen fibrillogenesis in predentin as well as altered dentin mineralization and enamel formation. J Histochem Cytochem. 2006;54(5):525–37.
6. Kalamajski S, Oldberg Å. Homologous sequence in lumican and fibromodulin leucine-rich repeat 5-7 competes for collagen binding. J Biol Chem. 2009;284(1):534–9.
7. Lauder RM, Huckerby TN, Nieduszynski IA. The structure of the keratan sulphate chains attached to fibromodulin isolated from articular cartilage. Eur J Biochem. 1996;242(2):402–9.
8. Zheng Z, Granado HS, Li C. Fibromodulin, a multifunctional matricellular modulator. J Dent Res. 2023;102(2):125–34.
9. Gori F, Schipani E, Demay MB. Fibromodulin is expressed by both chondrocytes and osteoblasts during fetal bone development. J Cell Biochem. 2001;82(1):46–57.
10. Sjöberg A, Onnerfjord P, Mörgelin M, Heinegård D, Blom AM. The extracellular matrix and inflammation: fibromodulin activates the classical pathway of complement by directly binding C1q. J Biol Chem. 2005;280(37):32301–8.
11. Kopp A, Hebecker M, Svobodová E, Józsi M. Factor h: a complement regulator in health and disease, and a mediator of cellular interactions. Biomol Ther. 2012;2(1):46–75.
12. Geng Y, McQuillan D, Roughley PJ. SLRP interaction can protect collagen fibrils from cleavage by collagenases. Matrix Biol. 2006;25(8):484–91.
13. Lin L, Jia Z, Zhao Y, Wu Y, Zhao X, Li Y, et al. Use of limiting dilution method for isolation of nucleus pulposus mesenchymal stem/progenitor cells and effects of plating density on biological characteristics and plasticity. Biomed Res Int. 2017;2017:9765843.
14. Li C, Ha P, Jiang W, Haveles CS, Zheng Z, Zou M. Fibromodulin – a new target of osteoarthritis management? Front Pharmacol. 2019;10:1475.
15. Rees SG, Hughes Wassell DT, Waddington RJ, Embery G. Interaction of bone proteoglycans and proteoglycan components with hydroxyapatite. Biochim Biophys Acta. 2001;1568(2):118–28.
16. Li CS, Yang P, Ting K, Aghaloo T, Lee S, Zhang Y, et al. Fibromodulin reprogrammed cells: a novel cell source for bone regeneration. Biomaterials. 2016;83:194–206.
17. Säämänen AM, Salminen HJ, Rantakokko AJ, Heinegård D, Vuorio EI. Murine fibromodulin: cDNA and genomic structure, and age-related expression and distribution in the knee joint. Biochem J. 2001;355(Pt 3):577–85.
18. Goldberg M, Marchadier A, Vidal C, Harichane Y, Kamoun-Goldrat A, Kellermann O, et al. Differential effects of fibromodulin deficiency on mouse mandibular bones and teeth: a micro-CT time course study. Cells Tissues Organs. 2011;194(2–4):205–10.
19. Qian H, Xiao Y, Bartold PM. Immunohistochemical localization and expression of fibromodulin in adult rat periodontium and inflamed human gingiva. Oral Dis. 2004;10(4):233–9.
20. Matias MA, Li H, Young WG, Bartold PM. Immunohistochemical localization of fibromodulin in the periodontium during cementogenesis and root formation in the rat molar. J Periodontal Res. 2003;38(5):502–7.
21. Svensson L, Aszódi A, Reinholt FP, Fässler R, Heinegård D, Oldberg A. Fibromodulin-null mice have abnormal collagen fibrils, tissue organization, and altered lumican deposition in tendon. J Biol Chem. 1999;274(14):9636–47.

22. Wang L, Foster BL, Kram V, Nociti FH, Zerfas PM, Tran AB, et al. Fibromodulin and biglycan modulate periodontium through TGFβ/BMP signaling. J Dent Res. 2014;93(8):780–7.
23. Goldberg M, Smith AJ. Cells and extracellular matrices of dentin and pulp: a biological basis for repair and tissue engineering. Crit Rev Oral Biol Med. 2004;15(1):13–27.
24. Lipman K, Wang C, Ting K, Soo C, Zheng Z. Tendinopathy: injury, repair, and current exploration. Drug Des Devel Ther. 2018;12:591–603.
25. Bi Y, Ehirchiou D, Kilts TM, Inkson CA, Embree MC, Sonoyama W, et al. Identification of tendon stem/progenitor cells and the role of the extracellular matrix in their niche. Nat Med. 2007;13(10):1219–27.
26. Xu X, Zhang Y, Ha P, Chen Y, Li C, Yen E, et al. A novel injectable fibromodulin-releasing granular hydrogel for tendon healing and functional recovery. Bioeng Transl Med. 2023;8(1):e10355.
27. Bressler HB, Markus M, Bressler RP, Friedman SN, Friedman L. Temporal tendinosis: a cause of chronic orofacial pain. Curr Pain Headache Rep. 2020;24(5):18.
28. Dupuis LE, Doucette L, Rice AK, Lancaster AE, Berger MG, Chakravarti S, et al. Development of myotendinous-like junctions that anchor cardiac valves requires fibromodulin and lumican. Dev Dyn. 2016;245(10):1029–42.
29. Lee EJ, Jan AT, Baig MH, Ashraf JM, Nahm SS, Kim YW, et al. Fibromodulin: a master regulator of myostatin controlling progression of satellite cells through a myogenic program. FASEB J. 2016;30(8):2708–19.
30. Andenæs K, Lunde IG, Mohammadzadeh N, Dahl CP, Aronsen JM, Strand ME, et al. The extracellular matrix proteoglycan fibromodulin is upregulated in clinical and experimental heart failure and affects cardiac remodeling. PLoS One. 2018;13(7):e0201422.
31. Lee EJ, Ahmad SS, Lim JH, Ahmad K, Shaikh S, Lee YS, et al. Interaction of fibromodulin and myostatin to regulate skeletal muscle aging: an opposite regulation in muscle aging, diabetes, and intracellular lipid accumulation. Cells. 2021;10(8):2083.
32. Lee EJ, Jan AT, Baig MH, Ahmad K, Malik A, Rabbani G, et al. Fibromodulin and regulation of the intricate balance between myoblast differentiation to myocytes or adipocyte-like cells. FASEB J. 2018;32(2):768–81.
33. Zheng Z, Zhang X, Dang C, Beanes S, Chang GX, Chen Y, et al. Fibromodulin is essential for fetal-type scarless cutaneous wound healing. Am J Pathol. 2016;186(11):2824–32.
34. Zheng Z, Lee KS, Zhang X, Nguyen C, Hsu C, Wang JZ, et al. Fibromodulin-deficiency alters temporospatial expression patterns of transforming growth factor-**β** ligands and receptors during adult mouse skin wound healing. PLoS One. 2014;9(6):e90817.
35. Jian J, Zheng Z, Zhang K, Rackohn TM, Hsu C, Levin A, et al. Fibromodulin promoted in vitro and in vivo angiogenesis. Biochem Biophys Res Commun. 2013;436(3):530–5.
36. Knoedler S, Broichhausen S, Guo R, Dai R, Knoedler L, Kauke-Navarro M, et al. Fibroblasts – the cellular choreographers of wound healing. Front Immunol. 2023;14:1233800.
37. Zheng Z, James AW, Li C, Jiang W, Wang JZ, Chang GX, et al. Fibromodulin reduces scar formation in adult cutaneous wounds by eliciting a fetal-like phenotype. Signal Transduct Target Ther. 2017;2:17050.
38. Khan FU, Owusu-Tieku NYG, Dai X, Liu K, Wu Y, Tsai HI, et al. Wnt/**β**-catenin pathway-regulated fibromodulin expression is crucial for breast cancer metastasis and inhibited by aspirin. Front Pharmacol. 2019;10:1308.
39. Chen M, Ba H, Lu C, Dai J, Sun J. Glial cell line-derived neurotrophic factor (GDNF) promotes angiogenesis through the demethylation of the fibromodulin (FMOD) promoter in glioblastoma. Med Sci Monit. 2018;24:6137–43.
40. Mayr C, Bund D, Schlee M, Moosmann A, Kofler DM, Hallek M, et al. Fibromodulin as a novel tumor-associated antigen (TAA) in chronic lymphocytic leukemia (CLL), which allows expansion of specific CD8+ autologous T lymphocytes. Blood. 2005;105(4):1566–73.
41. Pourhanifeh MH, Mohammadi R, Noruzi S, Hosseini SA, Fanoudi S, Mohamadi Y, et al. The role of fibromodulin in cancer pathogenesis: implications for diagnosis and therapy. Cancer Cell Int. 2019;19:157.

42. Raouf A, Ganss B, McMahon C, Vary C, Roughley PJ, Seth A. Lumican is a major proteoglycan component of the bone matrix. Matrix Biol. 2002;21(4):361–7.
43. Chakravarti S, Magnuson T, Lass JH, Jepsen KJ, LaMantia C, Carroll H. Lumican regulates collagen fibril assembly: skin fragility and corneal opacity in the absence of lumican. J Cell Biol. 1998;141(5):1277–86.
44. Giatagana EM, Berdiaki A, Tsatsakis A, Tzanakakis GN, Nikitovic D. Lumican in carcinogenesis – revisited. Biomolecules. 2021;11(9):1319.
45. Karamanou K, Franchi M, Proult I, Rivet R, Vynios D, Brézillon S. Lumican inhibits in vivo melanoma metastasis by altering matrix-effectors and invadopodia markers. Cells. 2021;10(4):841.
46. Lee YS, Park SJ, Lee JY, Choi E, Kim BJ. Benefits of lumican on human bone health: clinical evidence using bone marrow aspirates. Korean J Intern Med. 2022;37(4):821–9.
47. Krishnan A, Li X, Kao WY, Viker K, Butters K, Masuoka H, et al. Lumican, an extracellular matrix proteoglycan, is a novel requisite for hepatic fibrosis. Lab Investig. 2012;92(12):1712–25.
48. Schaefer L. Small leucine-rich proteoglycans in kidney disease. J Am Soc Nephrol. 2011;22(7):1200–7.
49. Bülow RD, Boor P. Extracellular matrix in kidney fibrosis: more than just a scaffold. J Histochem Cytochem. 2019;67(9):643–61.
50. Ahmad A, Ahmad R. Understanding the mechanism of hepatic fibrosis and potential therapeutic approaches. Saudi J Gastroenterol. 2012;18(3):155–67.
51. Feng S, Gao Y, Yin D, Lv L, Wen Y, Li Z, et al. Identification of lumican and fibromodulin as hub genes associated with accumulation of extracellular matrix in diabetic nephropathy. Kidney Blood Press Res. 2021;46(3):275–85.
52. Engebretsen KV, Lunde IG, Strand ME, Waehre A, Sjaastad I, Marstein HS, et al. Lumican is increased in experimental and clinical heart failure, and its production by cardiac fibroblasts is induced by mechanical and proinflammatory stimuli. FEBS J. 2013;280(10):2382–98.
53. Zhang X, Cui S, Ding Y, Li Y, Wu B, Gao J, et al. Downregulation of B4GALT5 attenuates cardiac fibrosis through lumican and Akt/GSK-3β/β-catenin pathway. Eur J Pharmacol. 2024;963:176263.
54. Han Y, Li Z, Wu Q, Liu H, Sun Z, Wu Y, et al. B4GALT5 high expression associated with poor prognosis of hepatocellular carcinoma. BMC Cancer. 2022;22(1):392.
55. Breslow DK, Weissman JS. Membranes in balance: mechanisms of sphingolipid homeostasis. Mol Cell. 2010;40(2):267–79.
56. Kao WW, Liu CY. Roles of lumican and keratocan on corneal transparency. Glycoconj J. 2002;19(4–5):275–85.
57. Wang IJ, Chiang TH, Shih YF, Hsiao CK, Lu SC, Hou YC, et al. The association of single nucleotide polymorphisms in the 5′-regulatory region of the lumican gene with susceptibility to high myopia in Taiwan. Mol Vis. 2006;12:852–7.
58. Cai XB, Shen SR, Chen DF, Zhang Q, Jin ZB. An overview of myopia genetics. Exp Eye Res. 2019;188:107778.
59. Tsui MC, Liu HY, Chu HS, Chen WL, Hu FR, Kao WW, et al. The versatile roles of lumican in eye diseases: a review. Ocul Surf. 2023;29:388–97.
60. Chang WC, Lee CH, Chiou SH, Liao CC, Cheng CW. Aqueous lumican correlates with central retinal thickness in patients with idiopathic epiretinal membrane: a proteome study. Dis Markers. 2022;2022:9886846.
61. Itoh Y, Sahni V, Shnider SJ, McKee H, Macklis JD. Inter-axonal molecular crosstalk via Lumican proteoglycan sculpts murine cervical corticospinal innervation by distinct subpopulations. Cell Rep. 2023;42(3):112182.
62. Bouhrira N, DeOre BJ, Tran KA, Galie PA. Transcriptomic analysis of a 3D blood-brain barrier model exposed to disturbed fluid flow. Fluids Barriers CNS. 2022;19(1):94.
63. Colon-Caraballo M, Lee N, Nallasamy S, Myers K, Hudson D, Iozzo RV, et al. Novel regulatory roles of small leucine-rich proteoglycans in remodeling of the uterine cervix in pregnancy. Matrix Biol. 2022;105:53–71.

64. Liu C, Hu Y, Wang Z, Pan H, Ren Y, Li X, et al. The downregulation of placental lumican promotes the progression of preeclampsia. Reprod Sci. 2021;28(11):3147–54.
65. Salgado RM, Favaro RR, Zorn TM. Modulation of small leucine-rich proteoglycans (SLRPs) expression in the mouse uterus by estradiol and progesterone. Reprod Biol Endocrinol. 2011;9:22.
66. Kedem A, Ulanenko-Shenkar K, Yung Y, Youngster M, Avraham S, Yerushalmi GM, et al. The involvement of lumican in human ovulatory processes. Reprod Sci. 2022;29(2):366–73.
67. Simões RS, Soares JM, Simões MJ, Nader HB, Baracat MCP, Maciel GAR, et al. Small leucine-rich proteoglycans (SLRPs) in the endometrium of polycystic ovary syndrome women: a pilot study. J Ovarian Res. 2017;10(1):54.
68. Smith MM, Melrose J. Lumican, a multifunctional cell instructive biomarker proteoglycan has novel roles as a marker of the hypercoagulative state of long Covid disease. Int J Mol Sci. 2024;25(5):2825.
69. Chen L, Liao J, Klineberg E, Leung VY, Huang S. Small leucine-rich proteoglycans (SLRPs): characteristics and function in the intervertebral disc. J Tissue Eng Regen Med. 2017;11(3):602–8.
70. Li Z, Sun C, Chen M, Wang B. Lumican silencing alleviates tumor necrosis factor-**α**-induced nucleus pulposus cell inflammation and senescence by inhibiting apoptosis signal regulating kinase 1/p38 signaling pathway via inactivating Fas ligand expression. Bioengineered. 2021;12(1):6891–901.
71. Lee JY, Kim DA, Kim EY, Chang EJ, Park SJ, Kim BJ. Lumican inhibits osteoclastogenesis and bone resorption by suppressing Akt activity. Int J Mol Sci. 2021;22(9):4717.
72. Xiao D, Liang T, Zhuang Z, He R, Ren J, Jiang S, et al. Lumican promotes joint fibrosis through TGF-**β** signaling. FEBS Open Bio. 2020;10(11):2478–88.
73. Cho HJ, Lee YS, Kim DA, Moon SA, Lee SE, Lee SH, et al. Lumican, an exerkine, protects against skeletal muscle loss. Int J Mol Sci. 2022;23(17):10031.
74. Kashyap MK, Marimuthu A, Peri S, Kumar GS, Jacob HK, Prasad TS, et al. Overexpression of periostin and lumican in esophageal squamous cell carcinoma. Cancers (Basel). 2010;2(1):133–42.
75. Matsuda Y, Yamamoto T, Kudo M, Kawahara K, Kawamoto M, Nakajima Y, et al. Expression and roles of lumican in lung adenocarcinoma and squamous cell carcinoma. Int J Oncol. 2008;33(6):1177–85.
76. Karamanou K, Franchi M, Piperigkou Z, Perreau C, Maquart FX, Vynios DH, et al. Lumican effectively regulates the estrogen receptors-associated functional properties of breast cancer cells, expression of matrix effectors and epithelial-to-mesenchymal transition. Sci Rep. 2017;7:45138.
77. Nikitovic D, Berdiaki A, Zafiropoulos A, Katonis P, Tsatsakis A, Karamanos NK, et al. Lumican expression is positively correlated with the differentiation and negatively with the growth of human osteosarcoma cells. FEBS J. 2008;275(2):350–61.
78. Radwanska A, Litwin M, Nowak D, Baczynska D, Wegrowski Y, Maquart FX, et al. Overexpression of lumican affects the migration of human colon cancer cells through up-regulation of gelsolin and filamentous actin reorganization. Exp Cell Res. 2012;318(18):2312–23.
79. Yamamoto T, Matsuda Y, Kawahara K, Ishiwata T, Naito Z. Secreted 70kDa lumican stimulates growth and inhibits invasion of human pancreatic cancer. Cancer Lett. 2012;320(1):31–9.
80. Chen YC, Sosnoski DM, Mastro AM. Breast cancer metastasis to the bone: mechanisms of bone loss. Breast Cancer Res. 2010;12(6):215.
81. Sun C, Chen Y, Kim NH, Lowe S, Ma S, Zhou Z, et al. Identification and verification of potential biomarkers in gastric cancer by integrated bioinformatic analysis. Front Genet. 2022;13:911740.
82. Coulson-Thomas VJ, Coulson-Thomas YM, Gesteira TF, Andrade de Paula CA, Carneiro CR, Ortiz V, et al. Lumican expression, localization and antitumor activity in prostate cancer. Exp Cell Res. 2013;319(7):967–81.

83. Jeanne A, Untereiner V, Perreau C, Proult I, Gobinet C, Boulagnon-Rombi C, et al. Lumican delays melanoma growth in mice and drives tumor molecular assembly as well as response to matrix-targeted TAX2 therapeutic peptide. Sci Rep. 2017;7(1):7700.
84. Matheson S, Larjava H, Häkkinen L. Distinctive localization and function for lumican, fibromodulin and decorin to regulate collagen fibril organization in periodontal tissues. J Periodontal Res. 2005;40(4):312–24.
85. Wendel M, Sommarin Y, Heinegård D. Bone matrix proteins: isolation and characterization of a novel cell-binding keratan sulfate proteoglycan (osteoadherin) from bovine bone. J Cell Biol. 1998;141(3):839–47.
86. Tashima T, Nagatoishi S, Sagara H, Ohnuma S, Tsumoto K. Osteomodulin regulates diameter and alters shape of collagen fibrils. Biochem Biophys Res Commun. 2015;463(3):292–6.
87. Lin W, Gao L, Jiang W, Niu C, Yuan K, Hu X, et al. The role of osteomodulin on osteo/odontogenic differentiation in human dental pulp stem cells. BMC Oral Health. 2019;19(1):22.
88. Edsberg LE, Crowgey EL, Osborn PM, Wyffels JT. A survey of proteomic biomarkers for heterotopic ossification in blood serum. J Orthop Surg Res. 2017;12(1):69.
89. Bianco P, Fisher LW, Young MF, Termine JD, Robey PG. Expression of bone sialoprotein (BSP) in developing human tissues. Calcif Tissue Int. 1991;49(6):421–6.
90. Zappia J, Tong Q, Van der Cruyssen R, Cornelis FMF, Lambert C, Pinto Coelho T, et al. Osteomodulin downregulation is associated with osteoarthritis development. Bone Res. 2023;11(1):49.
91. Pesesse L, Sanchez C, Walsh DA, Delcour JP, Baudouin C, Msika P, et al. Bone sialoprotein as a potential key factor implicated in the pathophysiology of osteoarthritis. Osteoarthr Cartil. 2014;22(4):547–56.
92. Ramstad VE, Franzén A, Heinegård D, Wendel M, Reinholt FP. Ultrastructural distribution of osteoadherin in rat bone shows a pattern similar to that of bone sialoprotein. Calcif Tissue Int. 2003;72(1):57–64.
93. Zhao W, von Kroge S, Jadzic J, Milovanovic P, Sihota P, Luther J, et al. Osteomodulin deficiency in mice causes a specific reduction of transversal cortical bone size. J Bone Miner Res. 2024;39(7):1025–41.
94. Rehn AP, Chalk AM, Wendel M. Differential regulation of osteoadherin (OSAD) by TGF-beta1 and BMP-2. Biochem Biophys Res Commun. 2006;349(3):1057–64.
95. Bengtsson E, Neame PJ, Heinegård D, Sommarin Y. The primary structure of a basic leucine-rich repeat protein, PRELP, found in connective tissues. J Biol Chem. 1995;270(43):25639–44.
96. Bengtsson E, Mörgelin M, Sasaki T, Timpl R, Heinegård D, Aspberg A. The leucine-rich repeat protein PRELP binds perlecan and collagens and may function as a basement membrane anchor. J Biol Chem. 2002;277(17):15061–8.
97. Grover J, Chen XN, Korenberg JR, Recklies AD, Roughley PJ. The gene organization, chromosome location, and expression of a 55-kDa matrix protein (PRELP) of human articular cartilage. Genomics. 1996;38(2):109–17.
98. Happonen KE, Fürst CM, Saxne T, Heinegård D, Blom AM. PRELP protein inhibits the formation of the complement membrane attack complex. J Biol Chem. 2012;287(11):8092–100.
99. Rucci N, Rufo A, Alamanou M, Capulli M, Del Fattore A, Ahrman E, et al. The glycosaminoglycan-binding domain of PRELP acts as a cell type-specific NF-kappaB inhibitor that impairs osteoclastogenesis. J Cell Biol. 2009;187(5):669–83.
100. Grover J, Lee ER, Mounkes LC, Stewart CL, Roughley PJ. The consequence of PRELP overexpression on skin. Matrix Biol. 2007;26(2):140–3.
101. Merle NS, Noe R, Halbwachs-Mecarelli L, Fremeaux-Bacchi V, Roumenina LT. Complement system part II: role in immunity. Front Immunol. 2015;6:257.
102. Williams AS, Mizuno M, Richards PJ, Holt DS, Morgan BP. Deletion of the gene encoding CD59a in mice increases disease severity in a murine model of rheumatoid arthritis. Arthritis Rheum. 2004;50(9):3035–44.
103. Bengtsson E, Aspberg A, Heinegard D, Sommarin Y, Spillmann D. The amino-terminal part of PRELP binds to heparin and heparan sulfate. J Biol Chem. 2000;275(52):40695–702.

104. Futaki S, Nakase I, Suzuki T, Youjun Z, Sugiura Y. Translocation of branched-chain arginine peptides through cell membranes: flexibility in the spatial disposition of positive charges in membrane-permeable peptides. Biochemistry. 2002;41(25):7925–30.
105. Grover J, Roughley PJ. Characterization of the human proline/arginine-rich end leucine-rich repeat protein (PRELP) gene promoter and identification of a repressor element. Biochem J. 1998;336((Pt 1)(Pt 1)):77–82.
106. Corpuz LM, Funderburgh JL, Funderburgh ML, Bottomley GS, Prakash S, Conrad GW. Molecular cloning and tissue distribution of keratocan. Bovine corneal keratan sulfate proteoglycan 37A. J Biol Chem. 1996;271(16):9759–63.
107. Tasheva ES, Funderburgh JL, Funderburgh ML, Corpuz LM, Conrad GW. Structure and sequence of the gene encoding human keratocan. DNA Seq. 1999;10(1):67–74.
108. Pellegata NS, Dieguez-Lucena JL, Joensuu T, Lau S, Montgomery KT, Krahe R, et al. Mutations in KERA, encoding keratocan, cause cornea plana. Nat Genet. 2000;25(1):91–5.
109. Nischal KK. Genetics of congenital corneal opacification—impact on diagnosis and treatment. Cornea. 2015;34(Suppl 10):S24–34.
110. Tahvanainen E, Forsius H, Kolehmainen J, Damsten M, Fellman J, de la Chapelle A. The genetics of cornea plana congenita. J Med Genet. 1996;33(2):116–9.
111. Eriksson AW, Lehmann W, Forsius J. Congenital cornea plana in Finland. Clin Genet. 1973;4(4):301–10.
112. Tahvanainen E, Forsius H, Karila E, Ranta S, Eerola M, Weissenbach J, et al. Cornea plana congenita gene assigned to the long arm of chromosome 12 by linkage analysis. Genomics. 1995;26(2):290–3.
113. Danielson KG, Fazzio A, Cohen I, Cannizzaro LA, Eichstetter I, Iozzo RV. The human decorin gene: intron-exon organization, discovery of two alternatively spliced exons in the 5′ untranslated region, and mapping of the gene to chromosome 12q23. Genomics. 1993;15(1):146–60.
114. Grover J, Chen XN, Korenberg JR, Roughley PJ. The human lumican gene. Organization, chromosomal location, and expression in articular cartilage. J Biol Chem. 1995;270(37):21942–9.
115. Igwe JC, Gao Q, Kizivat T, Kao WW, Kalajzic I. Keratocan is expressed by osteoblasts and can modulate osteogenic differentiation. Connect Tissue Res. 2011;52(5):401–7.
116. Melrose J, Fuller ES, Roughley PJ, Smith MM, Kerr B, Hughes CE, et al. Fragmentation of decorin, biglycan, lumican and keratocan is elevated in degenerate human meniscus, knee and hip articular cartilages compared with age-matched macroscopically normal and control tissues. Arthritis Res Ther. 2008;10(4):R79.
117. Garnero P. Use of biochemical markers to study and follow patients with osteoarthritis. Curr Rheumatol Rep. 2006;8(1):37–44.

Regulatory Role and Significance of Class III SLRPs: Osteoglycin, Epiphycan, and Opticin

5

Mamta Srinivasan

Abstract

Class III SLRP members are opticin, osteoglycin (also known as mimecan), and epiphycan, and they play a key role in regulating collagen fibrillogenesis. Recent evidence also suggests these protein can compensate for one another in the knockout mice model under various pathological conditions, suggesting functional redundancy. Notably Class III SLRP members possess the fewest leucine-rich repeats among the SLRP family and are frequently found as proteolytically cleaved fragmented forms. This chapter explores the individual roles of each Class III SLRP member in detail.

Keywords

Small leucine rich proteoglycans · Classification · Regulation · Disease · Cells · Tissues

5.1 Opticin (OPTC)

Opticin (OPTC), also called "oculoglycan," is encoded by the OPTC gene, localized to chromosome 1q31–q32. OPTC is a glycosylated protein, where an O-linked oligosaccharide is covalently attached to its protein core at the reactive amino (NH_2) terminal and is secreted into the extracellular matrix of various tissues.

Opticin, as the name suggests, is predominantly associated with the ocular (eye) tissues that include the cornea, iris, ciliary body, vitreous humor, choroid, and the

M. Srinivasan (✉)
Medical University of Americas, Nevis, St. Kitts and Nevis, West Indies
e-mail: m.srinivasan@mua.edu

P. Lama (ed.), *The Biological Role of Small Leucine-Rich Proteoglycans (SLRPs) in Matrix Homeostasis and Diseases*,
https://doi.org/10.1007/978-981-96-7890-7_5

retina. It was first extracted from the surface of bovine vitreous humor containing collagen fibrils, and recent evidence indicates that it binds to growth hormone to modulate its function. As a Class III SLRP member, it is characterized by the unique presence of tandem seven leucine-rich repeats (as opposed to twelve) flanked by cysteine clusters that do not contain a glycosaminoglycan side chain; therefore, opticin is also often considered as a "glycoprotein" with an amino terminal region, showing a cluster of sialylated O-linked oligosaccharides. Opticin is thus an exceptional member of the SLRP family bearing common functions associated with collagen fibril assembly and spacing, including its role in cell growth and proliferation [1]. It binds to collagen fibrils particularly between collagens II/XI or II/IX/XI, commonly found in cartilage and in the vitreous humor. This interaction modulates cell-matrix changes, and provides protection against proteases [2].

5.1.1 Opticin Glycosylation and Sulfation

Opticin is produced by chondrocytes, articular cells, synoviocytes, and osteoblasts. Glycosylation of OPTC modifies tissue distribution and functions, and it has two differentially O-glycosylated isoforms; and tyrosine sulfation in the retinal pigmented epithelium possibly altering antiangiogenic properties, including interaction with collagen and/or retinal growth hormone; thus, lack of opticin leads to disorganization of photoreceptor structure in the retina [3]. With complement factor H (CFH) and vitronectin also being sulfated in the retinal epithelium, and this interaction could be key in modulating the collagen and growth factors [3].

The unglycosylated core protein is seen in the nucleus and endoplasmic reticulum (ER) of hematological malignancies, chronic myeloid leukemia (CLL), and mantle cell lymphoma (MCL), suggesting its role in the pathogenesis of these malignant transformations. The opticin gene, being close to fibromodulin (FOMD), shows cluster expression in hematological malignancies. When upregulated opticin interact with collagen, growth hormone, and heparin sulphate GAG side chain, activating certain genes that regulate cell anchorage and growth factor sequestration thereby modulating angiogenesis which is a key factor in tumorigenesis and metastasis [4].

5.1.2 Opticin in Ocular Diseases and Osteoarthritis

In the vitreous humor of the eye, extracellular OPTC protein has also been shown to be antiangiogenic as it binds to collagen and inhibits $\alpha2\beta1$ and $\alpha1\beta1$ integrin-mediated endothelial cell adhesion. The molecular weight of the normal glycosylated opticin core protein varies approximately between 45 and 62 kDa and undergoes degradation into smaller sizes as fragments by proteases, such as MMP-1, 2, 3, 7, 8, 9, 13, ADAMTS4, and ADAMTS5, contributing to the pathophysiology in ocular diseases and osteoarthritis.

Opticin has a role in preventing the age-related posterior vitreous detachment in humans regulating the shape and structure of collagen fibrils. Their predisposition to be degraded by MMPs could be a factor leading to rhegmatogenous retinal detachment and other ocular diseases [5]. Its expression in other ECM structures, like ligament and skin, makes it a candidate for hereditary forms of Ehlers-Danlos and Stickler syndrome with joint and skin laxity along with retinal degeneration and detachments [5].

Thus, newer investigations have identified the presence of OPTC it in non-ocular tissues, such as the brain, skin, ligament, and cartilage. In normal ocular tissues, opticin is suggested to act as a biological glue between the vitreous collagen fibrils and the internal limiting lamina (ILL) of the retina and eye lens; thus, it is suggested to act as a therapeutic target for the pharmacological intervention in posterior vitreous detachment of the eyes. Further, mutations in the opticin gene are associated with age-related macular degeneration, one of the commonest causes of blindness in the world. Opticin, synthesized by the posterior nonpigmented ciliary epithelium, binds to laminin and type IV collagen in the basement of the internal limiting membrane (ILM) and lens capsule and around the vitreous humor in the eye. Increasing concentration of opticin in the vitreous tissue maintains the avascularity required for normal vision, and in vivo studies have demonstrated to have a therapeutic potential in patients with diabetic retinopathies, vitreous detachments, and trauma and these changes associates with vitreous hemorrhage [6]. Increased secretion and accumulation of opticin in the vitreous humor may play a role in retinal development and could potentially serve as a biomarker for age-related macular degeneration (AMD). Opticin expression in breast cancer cell lines suppresses proliferation, increases apoptosis, while minimising migration, which has a beneficial therapeutic potential. Additionally, OTCN is also shown to be downregulated in ciliary body tumor cells compared to normal ciliary body epithelium [4].

The most abundant large proteoglycan is aggrecan in cartilage tissue; opticin regulate fibril-fibril interaction and protect them from proteolytic damage by collagenases, a matrix degrading enzyme. Opticin also serves as a substrate for several other MMPs and aggrecanases enzyme during osteoarthritis with the most preferential substrate indicated as MMP-7 exerting its proteolytic action in a time-dependent degradation [16, 2]. As a consequence, decreased level of intact opticin in osteoarthritis pave the way to neovascularization in the development of osteoarthritis with aging [3, 5]. Interestingly, OPTC deficiency may exert a protective effect, as in vivo studies have shown an increase in synovial membrane thickness, while having no significant impact on the subchondral bone [3].

Opticin also forms dimers with decorin (Class I SLRP member), and this unique three-dimensional stoichiometric ratio allows bivalent interaction between the SLRPs to enable affinity for ligand binding, which allows interactions with substrates, inhibitors, activators, signaling molecules, and even neurotransmitters.

5.2 Osteoglycin/Mimecan

Osteoglycin (OGN), also called mimecan or osteoinductive factor, was first isolated as a protein that was found to induce ectopic bone formation when inoculated into rat skeletal muscle. OGN is a single-copy gene located on the human 9q22.31

chromosome, and it produces three mRNA transcripts as a result of differential splicing inside the first translated exon. Osteoglycin has a unique cysteine-rich sequence with only six to ten leucine-rich repeats (LRRs), and the *N*-terminal sequence bears the keratan sulfate side chains that are attached to the LRR domains with numerous glycosylation sites. OGN has a core protein size between 34 and 40 kDa, and apart from bone tissue, it has also been extracted from a variety of other tissues, such as the cornea, skin, aorta, sclera, vagus nerve, and cartilage, and at a lower level, from the ovary, cerebellum, skeletal muscle, heart, and kidney. OGN is secreted into the extracellular matrix after post-translational modification, thus interacting with fibrillar collagens, growth factors, other matrix molecules, and cell membrane receptors. In bone, OGN regulates osteoblast (bone cells) differentiation and collagen fibrillogenesis and is highly expressed in cardiovascular diseases, connective tissue disorders, and cancer. This protein, with its diverse molecular structure, contributes to the varied functional activity in different tissues and could be a candidate marker for diagnosis, prognosis, and therapeutics.

5.2.1 Osteoglycin and Bone

Osteoglycin protein is involved in ectopic bone formations and in vivo studies show it facilitates bone cell-receptor activation with TGFβ-like bone morphogenetic proteins 2 and 3; this interaction between OGN and TGF-β pathways suggests accelerated synthesis of anabolic activities regulating the formation of bone. TGF-β produced by osteoblasts also regulates and maintains bone metabolism through type I membrane receptor that promotes the formation of dimers with the type Π receptor, stimulating Ser/Thr kinase activity and causing phosphorylation of Smad2 and Smad3; thus these molecules are considered to be important for osteoblastic bone formation.

Osteoglycin enhances osteoclastogenesis and suppresses the differentiation of premature osteoblasts by decreasing the levels of Runx2 and Osterix but enhances the phenotype and mineralization of mature osteoblasts [7]. Osteoglycin secretion is the downstream mediator of NPY signaling through Y1 receptors on early osteoblasts, which also serves to regulate insulin sensitivity to aid skeletal adaptation to energy/weight change. The suppressed osteoglycin secretion by bone cells during weight gain due to reduced NPY signaling represents a mechanism to diminish insulin action in order to suppress energy uptake in insulin target tissues [7].

OGN cDNA is highly expressed in mouse hypertrophic chondrocytes as well as cochlear tissues. ALK2 signaling decreases OGN, which pathologically converts muscle tissues to bone in fibrodysplasia ossificans progressiva, suggesting its anabolic effects in bone.

5.2.2 Osteoglycin in Glucose and Insulin Modulation

The skeleton is an endocrine organ contributing to energy homeostasis, insulin sensitivity, glucose cell transport and metabolism. Osteoglycin is critical for mediating the adaptive response in bone to changing energy balance. Osteoblastic Y1 receptor signalling modulates insulin secretion and action via osteoglycin, which is a downstream mediator of Y1 receptor signalling in early osteoblasts and is involved in communication between bone, energy, and glucose homeostasis. Osteoglycin acts to suppress bone formation and modulate whole body energy supplies by altering glucose uptake through changes in insulin secretion and sensitivity, as well as by altering food intake through the central nervous system signalling by the neuropeptide (NYP Y1) receptor pathway [7]. Early osteoblasts through deletion of the Y1 receptor results in downregulation of osteoglycin, associated with reduced glucose tolerance and pancreatic insulin secretion. This suggests osteoglycin is a novel downstream mediator of osteoblastic Y1 receptor essential for the control of glucose homeostasis. OGN, along with lipocalin-2 (LCN2) synthesized by the bone, is dynamically linked to energy homeostasis in humans. Osteoglycin has been observed to be lower in middle-aged males with lower BMI, age, percent body fat, and fasting glucose level while higher OGN was associated with increased insulin sensitivity. OGN levels not only seem to decrease post-insulin secretion but also decline as age advances, which supports the hypothesis that OGN may play a mediating role in energy homeostasis and cardiometabolic health in humans [8]. Higher serum level of OGN were also seen to be associated with greater postexercise insulin sensitivity [8]. Serum OGN and LCN2 are dynamically influenced by exercise and high-glucose ingestion in healthy adults. Recent investigations also show that high-glucose ingestion leads to a decrease in circulating OGN and LCN2, and these effects were largely mitigated 3-hours and 24-hours postexercise. The decrease in OGN in a state of positive energy balance with excess weight or adiposity could lead to a decrease in neuropeptide Y (NPY) signalling, suppressing osteoblast-mediated secretion of OGN. Decreased OGN decreases appetite and food intake and reduces insulin secretion and insulin action in skeletal muscle [9]. Osteoglycin-maltose binding protein complex may inhibit food intake due to associated upregulation of interleukin (IL)-1beta and IL-6 in the hypothalamus. Osteoglycin induces c-Fos expression in the arcuate nucleus (Arc) of the hypothalamus, which significantly overlaps with NPY expression.

Interestingly, the cleaved osteoglycin is lower in plaques from patients with diabetes type 2, and it has a correlation with markers of insulin resistance (triglycerides and triglycerides/high-density lipoprotein) [10, 11]. Type 2 DM patients with high levels of triglycerides and low high density lipoprotein-c (HDL-c) have been predicted as an indicator for insulin resistance, which also enhances the risk for coronary vascular disease (CVD). A TG/HDL-C ratio >2.5 is an indicator for the long-term cause of mortality [12]. Increased osteoglycin levels have been observed in vascular tissue samples and in the serum of patients with type 2 diabetes (T2D) with and without cardiovascular diseases compared to healthy controls. The higher the insulin resistance in T2D patients (fourth quartile of TG/HDL-C index) higher is the levels of circulating osteoglycin. Osteoglycin expression influences the glucose levels in patients with T2D and also plays an indirect role in atherosclerosis. Autotaxin lysophosphatidic acid initiates a regulatory effect in vascular smooth muscles, especially during intimal hyperplasia, amplifies monocyte migration, and enhances vascular smooth muscle cell proliferation (VSMC), a feature in atherosclerosis. A study on calcified human embryonic kidney 293 T cells (HEK293T) overexpressed OGN, which increases autotaxin (ATX), a protein involved in T2D, insulin resistance, and obesity. OGN also upregulates the expression of autotaxin (ATX) in ectonucleotide pyrophosphatase/phosphodiesterase family member 2 (*ENPP2*), in vascular smooth muscle cells within calcified in vitro conditions [13].

Osteoglycin can be used as a biomarker of insulin resistance for type 2 diabetes and could be indirectly involved in the development of atherosclerosis, thus they act as an indicator for potential CVD risks [11].

5.2.3 OGN in Cardiovascular, Atherosclerosis, and Immune Cells

Atherosclerosis evolves as a multifactorial disease involving endothelial cell damage, triggering the immune response leading to plaque formation, with subsequent plaque rupture or erosion, it becomes a key cause of cardiovascular complications leading to stroke, heart failure, transient ischemic attacks (TIA), and vascular dementias.

While the neutrophils and macrophages are a major component of atheroma, endothelial cells and vascular smooth muscle cells (VSMCs) play a significant role, as the VSMCs, in turn, induce the formation and migration of foam cells. In concurrence, plaque formation includes fibrillogenesis with extracellular matrix (ECM) collagens, elastin, glycoproteins, glycosaminoglycans, and proteoglycans, and especially osteoglycin (OGN) or mimecan [13]. Collagen synthesis and fibrillogenesis are the key in the scar tissue formation after myocardial infraction healing.

Lucas N.L. et al. study on osteoglycin null mice and cardiac samples from post-myocardial infarction patients has observed overexpression of osteoglycin in wild-type mice, resulting in enhanced up regulation of collagen type I and III thus maintaining the tensile strength in the infarcted area. Osteoglycin is shown to enhance the polymerization of collagen fibers and dense collagen mesh formation increasing the density of collagen, which is required for the reduction of adverse remodeling and aids improved cardiac function after post-ischemia. In contrast, the osteoglycin null mice shows poor collagen quality and quantity ...(9)... Both in vitro and in vivo studies from the plasma concentration of MI patients detect increased osteoglycin, which could be a biomarker for patients with heart failure due to myocardial infraction (MI) [13].

OGN, which acts as a component of the vascular matrix, is produced by vascular smooth muscle cells, are downregulated by pro-angiogenic factors, and has been associated with plaque vulnerability, affecting the plaque rupture and further complications. OGN expression is observed in both tunica media and adventitia of blood vessels, and their concentration increases 14 days after balloon injury [11, 14], which is in accordance with its expression in the cultured rat VSMCs with the cessation of VSMC proliferation [15]. Two weeks into the maximal VSMC proliferation and remodeling, mRNA expression of osteoglycin in neonatal rat aorta increases [14], while the in situ human coronary artery shows osteoglycin mRNA expression by normal medial VSMCs was downregulated in a subset of intimal

VSMCs, indicating their differentiation as observed by the contractile markers SM22 alpha [14].

Increased activity and proliferation of VSMC also correlates with α-actin fibers in atherosclerotic plaques from asymptomatic patients and such changes are inversely associated with lipid metabolism and active caspase-3 expressions, suggesting lower levels of smooth muscle cell apoptosis in vitro. Annexin V studies have identified that the apoptosis rate of HASMCs overexpressing mimecan was elevated compared with control cells [15]. Apoptosis of VSMC has been shown to promote calcification in vessels, which has a significant role in the pathogenesis of atherosclerosis reducing the thickness of neointima and fibrous capsule. Increased proportion of apoptosis in the VMSC contributes to unstable plaque as it decreases collagen synthesis, increasing the chances of dislodgement [15]. Additionally, lower levels of cleaved osteoglycin were observed in patients undergoing surgery within 1 month of symptoms compared to plaques from asymptomatic patients [10]. It is also reported that knockdown of OGN/mimecan enhances primary cultured VSMC proliferation induced by FBS or Angiotensin II [17]. Angiogenic factors, including cytokines, fibroblast growth factor-2, transforming growth factor-β, platelet-derived growth factor, Angiotensin II, and oncostatin M, inhibit mimecan/osteoglycin expression, while MCP-1 enhances the effect of these cytokines on OGN [14, 20]. The balance between the degradation and formation of ECM components is of great importance for plaque stability and the prognosis of atherosclerotic complications.

The inflammatory process in the formation of atherosclerosis induces the activation of matrix metalloproteinases (MMPs), which are synthesized by the VSMCs and macrophages. MMP 1, 2, 9, and 10 are shown to degrade the collagen and OGN in the tunica intima of blood vessels and facilitate the plaque formation. MMP-2 and MMP-9 facilitate the development of atherosclerosis via matrix disruption and VSMC migration, of which MMP-9 is essential for neointimal and arterial lesions [15]. Mimecan also correlates positively with levels of MMP-9 and MCP-1 in human atherosclerotic carotid plaques [18]. Decorin, another class I SLRP is overexpressed, and found to significantly increase MMP-2 production [15]. Osteoglycin, being structurally similar to decorin, could mimic the action by regulating cell migration and the phenotypic differentiation of VSMCs [15]. Cleaved osteoglycin is significantly higher in asymptomatic plaques and MMP-12 has been shown to cleave osteoglycin, in vitro, leading to a stable plaque formation, which lowers the risk of future cardiovascular complications [10]. While increased expression of OGN is observed in VSMC proliferation, apoptosis of VSMCs has a role in the development of atherosclerosis as it decreases the thickness of neointima, thus showing the expression and implication OGN in different stages of atherosclerosis.

Collagen maturation affects scar tissue formation in atherosclerosis and post-MI. Loss of mimecan has an effect on post-MI cardiac tissue, resulting in larger collagen fibrils [13]. Expression of mimican in circulating and resident cardiac macrophages is also increased in cardiac inflammation during viral myocarditis [19].

Glycosylation of osteoglycin has resulted in variants like dermatan sulfate glycosylated form, and dermatan sulfate-OGN (34 kDa) needed for collagen fibrillogenesis in the heart; O-linked chondroitin sulfate OGN variant (72-kDa) is present in the membrane of immune cells, where it binds and activates TLRs during inflammation [21]. Located in circulating and cardiac innate immune cells, OGN is specific for innate immunity, which binds to and also activates the TLR-4 signaling pathway [19]. Bacterial endotoxins and viral infections target TLR-4, which plays a role in initiating the innate immune response by two pathways; leading to the induction of pro-inflammatory cytokines, and the second pathway is activated by the adaptors TRIF (TIR-domain-containing adaptor protein inducing interferon-β) and TRAM (TRIF-related adaptor molecule), which induces type I interferons. This activation also increases IL-6, TNF-α, IL-1β, and IL-12 expression. It is also significant to note that OGN fragments is the result of O-linked glycosylation of the 32-kDa protein core of OGN. The absence of OGN decreases cytokine production by both circulating and cardiac leukocytes upon (systemic) LPS exposure and reduces cardiac inflammation and injury in viral myocarditis.

5.2.4 Osteoglycin and Cancer

In colorectal cancer, OGN acts as a tumor suppressor by enhancing T-lymphocyte recruitment with decreased expression of pro-angiogenic factors, such as VEGF, that are shown to repress proinflammatory upregulation of chemokines and activation of endothelial cells, enhancing T-lymphocyte functions. Thus, the ability of OGN to impede angiogenesis while promoting immune system activation, as the recruitment of T-lymphocytes, suggests that OGN is a potent onco-suppressive molecule in colorectal cancer, which can be explored for further therapeutic investigations in controlled conditions where OGN does not bind to other several signaling receptors [22]. OGN is found in a variety of cancer tissues and is implicated in cellular processes linked to tumorigenesis, including cell proliferation, invasion, metastasis, and epithelial-mesenchymal transition (EMT), by suppressing the EGFR/AKT/Zeb-1 axis and via PI3K/Akt/mTOR pathway [23]. Low OGN expression in tumor cells are more likely to metastasize as observed in the in vivo studies on ovarian cancer. The high OGN expression in ovarian, pancreatic, and hepatocellular carcinomas has shown to decrease invasive ability of hepatocellular carcinoma cells and reduced metastasis, possibly regulating the expression of MMPs, E-cadherin, and N-cadherin [24].

In pancreatic ductal adenocarcinoma, the most notable cancer-associated fibroblast (CAF) subtypes, myofibroblastic CAFs (myCAFs), inflammatory CAFs (iCAFs), and OGN are required for upregulation of fibroblastic activity and myofibroblast differentiation, contributing to tissue remodeling and limiting injury to pancreatic tissue. The OGN-positive cells on differentiation express contractile genes, such as *Acta2* and *Tagln*, in skin wounds, which could possibly lead to vascular remodeling.

5.3 Epiphycan

Epiphycan, also called PG-Lb, has an amino-terminal dermatan sulfate side chain and two O-linked oligosaccharides, which contain seven leucine-rich repeats (LRRs). The gene located on chromosome 12q21.33 is known to regulate fibrillogenesis by interacting with collagen fibrils and other extracellular matrix proteins. It was first isolated from chick/bovine fetal epiphyseal cartilage, thus its name epiphycan and PG-Lb, signifying that it is a proteoglycan. The human homolog dermatan sulfate proteoglycan-3 (DSPG3) encodes a predicted 322-amino-acid protein with three potential glycosaminoglycan attachment sites, two N-glycosylation sites, a polyglutamic acid stretch, and six cysteines. Its suggested expression is predominant in cartilage, ligament, as well as endometrium, placental tissues, and testis (its significance unknown). As a component of ECM, epiphycan has a role in cartilage development and joint integrity maintenance.

5.3.1 Epiphycan in Chondrogenesis and ECM Maintenance

EPYC primarily participates in extracellular matrix (ECM) organization, collagen binding, and ECM binding to glycosaminoglycan and integrin receptors. Epiphycan is expressed in a stage-dependent differentiation of chondrocytes in the epiphyseal cartilage during endochondral bone formation, interacting with collagen type II and IX, which indicates its role in chondrocyte maturation and maintenance of its integrity. Expressed in the proximal cartilage, and proliferating zone of the growth plate in early stages, along with high expression of *Col2a1*, it is significantly absent in the zone of hypertrophic cartilage and perichondrium. Expressed in all zones except hypertrophic chondrocytes of the epiphyseal plate, EPYC expression indicates spatial maturation during early endochondral bone growth [25]. Substantiated by the in vivo studies, epiphycan deficiency displays decreased size of long bones and early development of OA as age advances, with consistent collagen loss associated with an increase in metalloproteinases MMP-2, MMP-14 and serine protease HtrA1 [26]. Shorter femurs in epiphycan-deficient male may indicate it's a correlation with its expression/absence in the testis, where epiphycan can be a cartilage differentiation marker during early embryogenesis. Increased expression of EPYC is associated with cartilage degeneration as OA progresses. It modulates the activation of the IL-17 signaling pathway within chondrocytes by regulating the interaction between IL-17A and its receptor, IL-17RA. IL-1β stimulation augments secretion of IL-17A by synoviocytes and its accumulation at chondrocyte surface receptors, EPYC, enhances this secretion. Chemokine IL-17A secretion is dysregulated by EPYC, thus activating the IL-17 signalling pathway in chondrocytes during the process of OA progression [27]. It is known to inhibit the transcription of the proteoglycan aggrecan and collagen fiber Col2 while increasing the transcription levels of the proteases ADATS5 and MMP13 [27]. Another transcriptional activator or repressor, Kruppel-like factor 9 (KLF9), is expressed at a low level in OA synovial tissue, where EPYC is expressed at a higher level. While IL-7 necessitates oxidative stress,

Kruppel-like factor 9 (KLF9) also has a significant role in oxidative stress responses in the occurrence of OA. Expression of KLF9 in resting memory CD_4^+T cells, activated NK cells, and activated mast cells signifies an immunomodulatory response leading to increased KLF9-dependent ROS, causing cartilage degradation [28]. Thus, EPYC with significant differences in the expression levels in OA cartilage not only is a potential diagnostic marker but also a therapeutic target in the treatment of osteoarthritis [27], while varying levels of KLF9 could be an additional diagnostic marker, but their defined roles needs to be further explored [28].

5.3.2 Epiphycan in Cancer Signaling

The small seven leucine-rich repeat epiphycan expressed in the ECM has pathological significance in tumor progression and metastasis. EPYC has a role in ECM organization via the PI3K-Akt signalling pathway for focal adhesion, ECM-receptor interaction, protein digestion, and absorption.

Ovarian cancer (serous cystadenocarcinomas and mucinous cystadenocarcinomas), prostate cancer, and pancreatic cancer cells express epiphycan in high concentrations with a positive correlation of genes associated with PI3K-Akt signalling pathway, focal adhesion kinases, protein digestion and absorption, and ECM-receptor interaction, while the highest expression is observed in metastatic ovarian and pancreatic cancers [25, 29]. EPYC possibly also interacts with phospholipase Cg2 (PLCG2) and CRK proto-oncogene, also named adaptor protein (CRK, or p38, or CRKII), thus playing a critical role in various malignant cells [25]. As PLCG2 is highly expressed in hematopoietic cells responsible for regulating immune responses, platelet adhesion, and spreading, its association with EPYC correlates with the malignant behavior of cells, while CRK, a member of the adapter protein families, is known to be implicated in IGF-1 regulatory pathways. The ECM receptor interaction via EPYC and its expression with SNAIL-2 activate the TGF-β/PI3K/AKT signalling pathway, signifying the property of invasion and metastasis of ovarian cancer cells [30]. EPYC can be a candidate gene as a biomarker for malignancies and metastatic prognosis while being a potential therapeutic target.

References

1. Reardon AJ, Le Goff M, Briggs MD, McLeod D, Sheehan JK, Thornton DJ, Bishop PN. Identification in vitreous and molecular cloning of opticin, a novel member of the family of leucine-rich repeat proteins of the extracellular matrix. J Biol Chem. 2000;275(3):2123–9. https://doi.org/10.1074/jbc.275.3.2123. ISSN 0021-9258.
2. Hansen U, Holmes DF, Bruckner P, Bishop PN. Analysis of opticin binding to collagen fibrils identifies a single binding site in the gap region and a high specificity towards thin heterotypic fibrils containing collagens II, and XI or V/XI. PLoS One. 2020;15(8):e0234672. https://doi.org/10.1371/journal.pone.0234672.
3. Kanan Y, Siefert JC, Kinter M, Al-Ubaidi MR. Complement factor H, vitronectin, and opticin are tyrosine-sulfated proteins of the retinal pigment epithelium. PLoS One. 2014;9(8):e105409. https://doi.org/10.1371/journal.pone.0105409. PMID: 25136834; PMCID: PMC4138151.

4. Mikaelsson E, Österborg A, Tahmasebi Fard Z, et al. Opticin, a small leucine-rich proteoglycan, is uniquely expressed and translocated to the nucleus of chronic lymphocytic leukemia cells. Exp Hematol Oncol. 2013;2:23. https://doi.org/10.1186/2162-3619-2-23.
5. Tío L, Martel-Pelletier J, Pelletier J-P, Bishop PN, Roughley P, Farran A, Benito P, Monfort J. Characterization of opticin digestion by proteases involved in osteoarthritis development. Joint Bone Spine. 2014;81(2):137–41. https://doi.org/10.1016/j.jbspin.2013.05.007. ISSN 1297-319X.
6. del Amo EM, Griffiths JR, Klaska IP, Hoke J, White A, Aarons L, Cooper GJS, Bainbridge JWB, Bishop PN, Unwin RD. Mol Pharm. 2020;17(7):2390–7. https://doi.org/10.1021/acs.molpharmaceut.0c00151.
7. Lee NJ, Ali N, Zhang L, Qi Y, Clarke I, Enriquez RF, Brzozowska M, Lee IC, Rogers MJ, Laybutt DR, Center JR, Baldock PA, Herzog H. Osteoglycin, a novel coordinator of bone and glucose homeostasis. Mol Metab. 2018;13:30–44. https://doi.org/10.1016/j.molmet.2018.05.004. ISSN 2212-8778.
8. Bauer C, Tacey A, Garnham A, Smith C, Woessner MN, Lin X, Zarekookandeh N, Hare DL, Lewis JR, Parker L, Levinger I. The effects of acute high-intensity interval exercise and hyperinsulinemic-euglycemic clamp on osteoglycin levels in young and middle-aged men. JBMR Plus. 2022;6(11):e10667. https://doi.org/10.1002/jbm4.10667. PMID: 36398112; PMCID: PMC9664545.
9. Parker L, Ang T, Morrison DJ, Lee NJ, Levinger I, Keske MA. Prior aerobic exercise mitigates the decrease in serum osteoglycin and lipocalin-2 following high-glucose mixed-nutrient meal ingestion in young men. Am J Physiol Endocrinol Metab. 2022;323(3):E319–32. https://doi.org/10.1152/ajpendo.00025.2022. Epub 2022 June 29.
10. Al-Sharify D, Nielsen SH, Matthes F, Tengryd C, Sun J, Genovese F, Karsdal MA, Nilsson J, Goncalves I, Edsfeldt A. Increased proteolytic cleavage of osteoglycin is associated with a stable plaque phenotype and lower risk of cardiovascular events. Atherosclerosis. 2022;355:8–14.
11. González-Salvatierra S, García-Fontana B, Martínez-Heredia L, Lacal J, Andújar-Vera F, Sanabria-de la Torre R, Moratalla-Aranda E, Lozano-Alonso S, García-Fontana C, Muñoz-Torres M. Exploring the role of osteoglycin in type 2 diabetes: implications for insulin resistance and vascular pathophysiology. Am J Physiol Endocrinol Metab. 2023;325(5):E649–60. https://doi.org/10.1152/ajpendo.00320.2023. Epub 2023 Oct 11. PMID: 37819194; PMCID: PMC10874653.
12. Kosmas CE, Rodriguez Polanco S, Bousvarou MD, Papakonstantinou EJ, Peña Genao E, Guzman E, Kostara CE. The triglyceride/high-density lipoprotein cholesterol (TG/HDL-C) ratio as a risk marker for metabolic syndrome and cardiovascular disease. Diagnostics (Basel). 2023;13(5):929. https://doi.org/10.3390/diagnostics13050929. PMID: 36900073; PMCID: PMC10001260.
13. Van Aelst LN, Voss S, Carai P, Van Leeuwen R, Vanhoutte D, Sanders-van Wijk S, Eurlings L, Swinnen M, Verheyen FK, Verbeken E, Nef H, Troidl C, Cook SA, Brunner-La Rocca HP, Möllmann H, Papageorgiou AP, Heymans S. Osteoglycin prevents cardiac dilatation and dysfunction after myocardial infarction through infarct collagen strengthening. Circ Res. 2015;116(3):425–36. https://doi.org/10.1161/CIRCRESAHA.116.304599. Epub 2014 Dec 17.
14. Shanahan CM, Cary NR, Osbourn JK, Weissberg PL. Identification of osteoglycin as a component of the vascular matrix. Differential expression by vascular smooth muscle cells during neointima formation and in atherosclerotic plaques. Arterioscler Thromb Vasc Biol. 1997;17(11):2437–47. https://doi.org/10.1161/01.atv.17.11.2437.
15. Zhang HJ, Wang J, Liu HF, Zhang XN, Zhan M, Chen FL. Overexpression of mimecan in human aortic smooth muscle cells inhibits cell proliferation and enhances apoptosis and migration. Exp Ther Med. 2015;10(1):187–92. https://doi.org/10.3892/etm.2015.2444. Epub 2015 Apr 22. PMID: 26170933; PMCID: PMC4487010.
16. Monfort J, Tardif G, Roughley P, Reboul P, Boileau C, Bishop PN, Pelletier JP, Martel-Pelletier J. Identification of opticin, a member of the small leucine-rich repeat proteoglycan family, in human articular tissues: a novel target for MMP-13 in osteoarthritis. Osteoarthr Cartil. 2008;16(7):749–55. https://doi.org/10.1016/j.joca.2007.11.007.

17. Gu XS, Lei JP, Shi JB, Lian WL, Yang X, Zheng X, Qin YW. Mimecan is involved in aortic hypertrophy induced by sinoaortic denervation in rats. Mol Cell Biochem. 2011;352(1–2):309–16. https://doi.org/10.1007/s11010-011-0767-8. Epub 2011 Apr 26.
18. Tengryd C, Nielsen SH, Cavalera M, Bengtsson E, Genovese F, Karsdal M, Dunér P, Orho-Melander M, Nilsson J, Edsfeldt A, Gonçalves I. The proteoglycan mimecan is associated with carotid plaque vulnerability and increased risk of future cardiovascular death. Atherosclerosis. 2020;313:88–95. https://doi.org/10.1016/j.atherosclerosis.2020.09.011. Epub 2020 Sept 25.
19. Rienks M, Papageorgiou A, Wouters K, Verhesen W, Leeuwen RV, Carai P, Summer G, Westermann D, Heymans S. A novel 72-kDa leukocyte-derived osteoglycin enhances the activation of toll-like receptor 4 and exacerbates cardiac inflammation during viral myocarditis. Cell Mol Life Sci. 2017;74(8):1511–25. https://doi.org/10.1007/s00018-016-2423-7. Epub 2016 Nov 23. PMID: 27878326; PMCID: PMC5357299.
20. Kampmann A, Fernandez B, Deindl E, Kubin T, Pipp F, Eitenmuller I, Hoefer IE, Schaper W, Zimmermann R. The proteoglycan osteoglycin/mimecan is correlated with arteriogenesis. Mol Cell Biochem. 2009;322:15–23.
21. Deckx S, Heymans S, Papageorgiou A-P. The diverse functions of osteoglycin: a deceitful dwarf, or a master regulator of disease? FASEB J. 2016;30:2651–61. https://doi.org/10.1096/fj.201500096R.
22. Hu X, Li YQ, Li QG, et al. Osteoglycin (OGN) reverses epithelial to mesenchymal transition and invasiveness in colorectal cancer via EGFR/Akt pathway. J Exp Clin Cancer Res. 2018;37:41. https://doi.org/10.1186/s13046-018-0718-2.
23. Xu T, Zhang R, Dong M, Zhang Z, Li H, Zhan C, Li X. Osteoglycin (OGN) inhibits cell proliferation and invasiveness in breast cancer via PI3K/Akt/mTOR signaling pathway. Onco Targets Ther. 2019;12:10639–50.
24. Jing-min Y, Yang M-Y, Zhao X, Jing-hui H. Effects of osteoglycin (OGN) on the growth and metastatic ability of epithelial ovarian cancer. J Biol Regul Homeost Agents. 2023;37(1):455–62. https://doi.org/10.23812/j.biol.regul.homeost.agents.20233701.45.
25. Deng L, Wang D, Chen S, Hu W, Zhang R. Epiphycan predicts poor outcomes and promotes metastasis in ovarian cancer. Front Oncol. 2021;11:653782. https://doi.org/10.3389/fonc.2021.653782. PMID: 34888227; PMCID: PMC8650094.
26. Nuka S, Zhou W, Henry SP, Gendron CM, Schultz JB, Shinomura T, Johnson J, Wang Y, Keene DR, Ramírez-Solis R, Behringer RR, Young MF, Höök M. Phenotypic characterization of epiphycan-deficient and epiphycan/biglycan double-deficient mice. Osteoarthritis Cartilage. 2010;18(1):88–96. https://doi.org/10.1016/j.joca.2009.11.006. Epub 2009 Nov 18. PMID: 19932218; PMCID: PMC3013283.
27. Huang J, Bu Z, Liu W, Zhou Z, Hu J, Yu J, Wang H, Xu S, Wu P. Cartilage decellularized matrix hydrogel loaded with protocatechualdehyde for targeted epiphycan treatment of osteoarthritis. Mater Today Bio. 2024;27:101124. https://doi.org/10.1016/j.mtbio.2024.101124. PMID: 38994469; PMCID: PMC11237976.
28. Zhang J, Zhang S, Zhou Y, Qu Y, Hou T, Ge W, Zhang S. KLF9 and EPYC acting as feature genes for osteoarthritis and their association with immune infiltration. J Orthop Surg Res. 2022;17(1):365. https://doi.org/10.1186/s13018-022-03247-6. PMID: 35902862; PMCID: PMC9330685.
29. Parker L, Ang T, Morrison DJ, Lee NJ, Levinger I, Keske MA. Prior aerobic exercise mitigates the decrease in serum osteoglycin and lipocalin-2 following high-glucose mixed-nutrient meal ingestion in young men. Am J Physiol Endocrinol Metab. 2022;323(3): E319–E332. https://doi.org/10.1152/ajpendo.00025.2022. PMID: 35767699
30. Deng L, Wang D, Chen S, Hu W, Zhang R. Epiphycan Predicts Poor Outcomes and Promotes Metastasis in Ovarian Cancer. Front Oncol. 2021 Nov 23;11:653782. https://doi.org/10.3389/fonc.2021.653782. PMID: 34888227; PMCID: PMC8650094.

Regulatory Role and Significance of Class IV SLRPs: Tsukushi, Chondroadherin, and Nyctalopin

6

Polly Lama and Sonam Choden Bhutia

Abstract

The noncanonical class IV SLRPs chondroadherin, nyctalopin, and tsukushi share a common evolutionary origin, and these SLRPs provide mechanical strength, adhesion, and support to the cellular microenvironment that shapes tissue-specific function. At a molecular level, characteristic features of class IV SLRPs include the lack of ear repeats, and they present a protein core molecule with eleven leucine-rich repeats attached to the N-terminus. Nyctalopin is the only SLRP member with a glycosylphosphatidylinositol-anchored side chain attached to the C-terminus and linked to the X chromosome. Mutations in nyctalopin have been reported to cause congenital stationary night blindness (CSNB1). Chondroadherin is a cartilage growth plate-specific SLRP that is distinct from other members as it presents a double disulfide loop close to its C-terminus. Tsukushi has important regulatory functions in modulating BMP and Wnt signaling pathways in various types of tissue. Accordingly, this chapter aims to highlight the importance of Class IV SLRPs in regulating tissue-specific functions, particularly in the tissue types where they are predominantly expressed.

Keywords

Small leucine rich proteoglycans · Regulation · Bones · Cartilage · Eye · Hair

P. Lama (✉)
Department of Anatomy, Sikkim Manipal Institute of Medical Sciences, Sikkim Manipal University, Tadong, India
e-mail: polly.lama@smims.smu.edu.in

S. C. Bhutia
Department of Biochemistry, Sikkim Manipal Institute of Medical Sciences, Sikkim Manipal University, Tadong, India

P. Lama (ed.), *The Biological Role of Small Leucine-Rich Proteoglycans (SLRPs) in Matrix Homeostasis and Diseases*,
https://doi.org/10.1007/978-981-96-7890-7_6

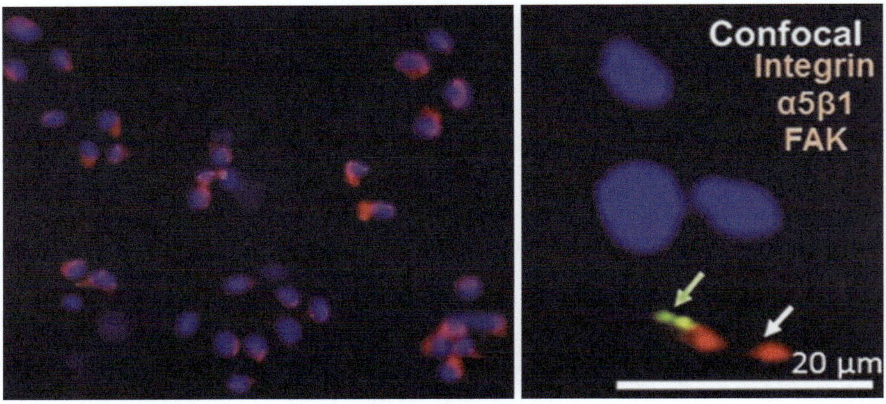

Fig. 6.1 Low-and high-resolution confocal laser scanning microscope images of disc cells expressing chondroadherin (left), integrin α5β1 receptor, along with focal adhesion kinase (right) in disc degeneration. Nuclei are stained blue with DAPI and chondroadherin and integrin receptors (α5β1) are stained red. Scale bar 200 and 20 µm. (Figure adapted from Dr. Polly Lama, Molecular Matrix Lab)

6.1 Chondroadherin (CHAD)

Chondroadherin is a collagen and integrin α2β1 receptor-binding SLRP expressed in cartilage and bone. Non-expression of CHAD in mice leads to thinner cortical bone and a longer proliferative growth plate zone [1]. CHAD regulates cell-matrix interactions through the α2β1 integrin receptors and also via cell surface proteoglycans, such as syndecans, providing specific signals to the cartilage cells. Additionally, they also act as an anchor to the matrix by binding tightly to collagen types I, II, and VI. CHAD is highly expressed in cartilaginous tissues (Fig. 6.1) and at lower levels it is detected in bone, tendon [2, 3], and eye [4]. In bovine bone, CHAD is implicated to directly interact with calcium phosphate mineral [5].

6.1.1 Regulatory Role of Chondroadherin in Cartilage

Regulatory role of chondroadherin in cartilage involves cell adhesion by activating ERK and MAPKs signaling pathway that does not facilitate cell spreading or division [6]. In humans, chondrosarcoma cells grown in collagen-chondroadherin substrate promotes cell spreading and proliferation, mediated by the integrin $\alpha_2\beta_1$ receptors with high expression levels of MAPK, as chondrosarcoma is a type of bone cancer that develops in cartilage cells [7]. Similarly, human articular cartilage chondrocytes grown on chondroadherin-coated surface promote adhesion and induce ERK phosphorylation with similar efficiency to the chondrosarcoma cell lines [8]. Bacterial vector expressing chondroadherin lacks posttranslational modifications, and CHAD cell binding to the matrix is similarly mediated by a straight peptide sequence without posttranslational modifications [9]. The binding of chondrocyte cells to chondroadherin is shown to be facilitated by the peptide sequence CQLRGLRRWLEAK, allowing adhesion, while other matrix proteins remain unaffected. Affinity purification of the peptide sequence by mass spectrometry allows separation and binding of molecular ligand via $\alpha_2\beta_1$ integrin receptor complex to the linear peptide sequence QLRGLRRWLEAK located in the cysteine loops at the C-terminus of chondroadherin [10] (Fig. 6.1). Thus, peptide sequence analysis suggests CHAD can play a role in matrix homeostasis by involving the regulation of growth factors, proteases, and inflammatory factors essential for the repair and adaptation of the cartilage matrix that is continuously exposed to different loading cycles and traumas.

6.1.2 Regulatory Role of Chondroadherin in Bone

Regulatory role of chondroadherin in bone involves maintenance of the(s) skeletal phenotype, as evidenced by mice with inactivated chondroadherin gene [2]. The null mice present distinct phenotypic changes in cartilage at 3–6 weeks in the epiphyseal growth plate with a widened proliferative zone, and around 4 months, these mice exhibit increased deposition of cartilage intermediate layer protein-1 and fibronectin, which is more pronounced in the female mice. Similarly, bone phenotypic alteration shows that one of the key markers for bone formation, bone sialoprotein mRNA-expressing cells in the proximal tibial metaphysis decreases while the osteoid deposition increases, indicating an alteration in mineral metabolism [2, 11]. Analysis of micro-CT reveals decreased cortical thickness but increased structure index of plates and rods composing the bone trabeculae. These changes accompany loss of function during mechanical testing, as the null mice femoral neck fails with decline in the tibial strength in 4 months [3]. Thus, change in skeletal phenotype shows the significance of chondroadherin in maintaining bone and cartilage tissue homeostasis without involving the longitudinal growth.

6.1.3 Regulatory Role of Chondroadherin in Ocular Tissue

Regulatory role of chondroadherin in ocular tissue includes developing and maintaining the structure of corneal epithelium, the endothelial cell layer, and type II collagen fibers to facilitate corneal hydration and transparency [12]. Absence of chondroadherin from corneal stroma shows loss in function related to cell adhesion and regulation of collagen fibrillogenesis, which could perhaps be due to the pericellular localization of chondroadherin that links to individual keratocytes [13]. The carbohydrate component of CHAD in ocular tissue carries a short oligosaccharide chain, which may interact with the proteoglycans keratan sulfate/chondroitin/dermatan sulfate of the corneal stroma required for interfibrillar spacing and tissue hydration [9]. The presence of CHAD in the ciliary body of eyes indicates a potential role in regulating the secretion of aqueous humour fluid, and expression of CHAD in the eye lens is suggested to maintain optical refraction. An abundance of

CHAD in the retinal tissue shows it is localized in the rod and cone cell layer, outer and inner plexiform cell layers, and possibly maintains the normal structure of the retina [6]. Increased decorin expression also occurs in the inner retinal cell layer of the transient ischemia disease model, where decorin aids the repair process in injured neural tissues. Thus, several retinal diseases are associated with mutations in the genes coding for SLRP family members or type II collagen to which CHAD is known to bind through the integrin $\alpha2\beta1$receptors, modulating cell-matrix interaction and functions that remain to be further determined.

6.2 Nyctalopin (*NYX*)

They are a membrane-bound SLRP expressed mainly in the retina and at a lower levels are also detected in the parts of brain, muscle, and kidney tissue, though their functional roles are not fully determined [14].

The nyctalopin gene is mutated in patients with X chromosome-linked congenital stationary night blindness (CSNB), and mutations in the type II collagen gene are associated with degenerative vitreoretinal changes in mice and Stickler syndrome in humans [15]. In cell culture analysis, nyctalopin is expressed on the cell membrane that is attached by either a glycosyl phosphatidylinositol (GPI) anchor or a transmembrane domain [16], and nyctalopin is the only SLRP member with an anchored side chain attached to the C-terminus and linked to the X chromosome.

6.2.1 Regulatory Role of Nyctalopin in the Retina

Regulatory role of nyctalopin in the retina is linked to human X-linked congenital stationary night blindness type-1 (CSNB-1) caused by mutations in the NYX gene encoding the protein [17]. CSNB-1 is characterized by loss of rod cell sensitivity with absence of the b-wave component in the dark- and light-adapted electroretinogram, sparing the a-wave [18]. The b-wave is generated by depolarizing bipolar cells of the retina to a light stimulus [19]. Therefore, the absence of the b-wave in CSNB1 patients shows a defect in synaptic transmission between photoreceptors and depolarizing bipolar cells. Mouse model with no b-wave carries a mutation in the NYX gene [20], representing CSNB-1. The GRM6 gene that encodes the expression of the postsynaptic metabotropic glutamate receptor (mGluR6) is also inactivated or mutated in mice and patients lacking the b-wave [21]. These similar phenotypic changes in CSNB-1 patients suggest the primary function of nyctalopin involves synaptic transmission between photoreceptors and depolarizing bipolar cells. Thus, mutations in the NYX gene have downstream consequences on the retinal photoreceptive function. It has also been shown that extracellular membrane-bound nyctalopin is required for the normal expression of transient receptor potential (TRP) channel, which binds the glutamate receptor 6 to the TRP melastatin-1 (TRPM-1) channel and transmits signals from photoreceptors to the dendrites of depolarizing bipolar cells in a mouse model [22]. Biochemical and gene expression analysis indicate that nyctalopin and TRPM-1 interact directly, and nyctalopin may even act as an accessory TRP channel critical for the proper localization of the visual synapses.

6.3 Tsukushi (*TSK*)

The protein TSK was discovered in the chick lens and is named after a Japanese horsetail plant due to its similarity in expression in the chick embryos [23]. TSK is a bone morphogenetic protein (BMP) antagonist, and details on the molecular structure of TSK are as described in Chaps. 1 and 2. In comparison to other SLRPs, tsukushi lacks the glycosaminoglycan side chains [24]. TSK is expressed during the early embryonic development and adulthood, and its function varies based on the organ it is expressed. Several organs and tissues, such as the brain, eye, ear, bone, liver, hair, and bloodexpress TSK [25–27]. However, functional analysis of TSK expression in these organ systems is not completely elucidated. Based on published results, the role of TSK is most documented in bone, ear, eyes, hair follicles, and wound healing.

6.3.1 Regulatory Role of Tsukushi in Bone

Regulatory role of tsukushi in bone is to coordinate multiple pathways by suppressing or increasing the individual activity of extracellular signaling molecules, such as BMP, Wnt, FGF, and TGF-β [28]. In vivo and in vitro data analyses show that tsukushi controls bone elongation and bone mass formation by involving growth plate chondrocytes and the process of chondrogenesis. TSK double knockout mice thus exhibit a reduction in weight and a short stature due to decreased longitudinal bone growth and loss of bone mass, and these phenotypic changes are observed in juvenile and adult knockouts [25]. Whole-mount β-gal staining of TSK knockout mice shows it is expressed in the humerus and costal cartilage with reduced thickness of the growth plate and abnormal shape of chondrocytes, suggesting skeletal abnormalities could be due to dysregulation of chondrogenic differentiation at the growth plate region. TSK, on the contrary, doesn't affect the size of the cranium as it is developed by intramembranous ossification, while long bones, such as the humerus, undergo intracartilaginous (endochondral) ossification. TSK deletion also results in significant downregulation of chondrogenesis marker genes, such as Sox-9 and Runx-2, that inhibit differentiation of chondrocytes in late stages of endochondral ossification controlled by multiple signaling pathways, such as Wnt, BMP, TGF, and FGF [29–31]. During chondrocyte differentiation, Wnt pathway suppresses the expression of Sox-9 while it activates and upregulates BMP, TGF, and FGF pathways, including the expression of collagen Xa (Col10a1) and matrix-degrading enzyme (MMP)-13.

6.3.2 Regulatory Role of Tsukushi in Hair Follicle

Regulatory role of tsukushi in hair follicle shows it is expressed in the hair bulb, bulge cells, and sebaceous gland throughout the hair cycle, including skin development around the follicles. TSK is related to cell differentiation and maintenance but deos not play a role in cell proliferation. Transforming growth factor-β (TGF-β), a signaling molecule, directly binds to TSK in the regulation of the hair follicle cycle, and targeted disruption of the TSK gene causes hair cycle growth delay with low levels of TGF-β1 and phosphorylated Smad2/3 (pSmad2/3) expressions. As human TSK is comparable to the mouse model, TSK knockout mice also show a delay in hair cycle, which downregulates the mRNA level of TGF-β signaling in mice. TGF-β1 null, and the knockout mice also show delayed catagen phase, which indicates the end of active hair growth, and injection of TGF-β1 into mouse skin, induced premature catagen development, suggesting TSK secreted from surface hair follicles bind to TGF-β1 outside the cell and perhaps controls cell regression at the catagen phase.

6.3.3 Regulatory Role of Tsukushi in Retinal Cells

Regulatory role of tsukushi in retinal cells effectively inhibits Wnt signaling in vivo and in vitro, and overexpression of TSK affects the retinal cells, ciliary body, and iris of the eyes [32]. TSK efficiently abrogates Wnt2b activity in differential response to retinal cells in vitro and in vivo, which is likely modulated by type I receptor low-density lipoprotein receptor-related protein and the seven-pass transmembrane protein frizzled-4 [33], containing a Wnt-binding site in the form of a cysteine-rich domain. Tsukushi can function as a Wnt antagonist in the developing eye, especially during peripheral differentiation of the eyes, regardless of the time, duration, and other experimental conditions applied. Furthermore, in the developing chick eyes, TSK is expressed in the ciliary body and iris epithelium, whereas Wnt2b is expressed in the adjacent anterior rim of the optic vesicle, regulating the differentiation of peripheral eye structures, such as the ciliary body and iris [23].

6.3.4 Regulatory Role of Tsukushi in Ears

Regulatory role of tsukushi in ears includes control of pathways that regulate inner ear development, where TSK interacts with Sox-2 and BMP-4 in stereocilia formation for the inner hair cells of ears [26]. TSK is detected in the pro-sensory and non-prosensory regions of the cochlear epithelial cells in embryonic and adult mice. TSK deficiency in the developing embryonic inner ear redistributes bone morphogenetic protein BMP-4 to form a thin layer throughout the cochlear epithelium, while BMP-4 depletion in outer sulcus region leads to slight accumulation in the pro-sensory regions [34]. In addition, loss of BMP signalling in the developing inner ear causes marked loss of hair cells, thus TSK knockout mice model show shortened stereocilia in the inner hair cells with no loss of inner or outer hair cells, however TSK knockout mice suffer from hearing loss even without the loss of the outer hair cells [35].

TSK loss suppresses Sox-2 gene at the pro-sensory regions, and Sox-2 is essential for upregulation of the transcription factor Atoh1 in both embryonic and postnatal cochlear progenitor cells, which triggers the differentiation of the hair cells [36]. Sox-2 deletion at late stages of the embryonic inner ear development elicits BMP-4 expression as well as the formation of ectopic inner hair cells. Thus, as a corollary, BMPs can repress Sox-2 posttranscriptionally.

References

1. Tillgren V, Ho JC, Önnerfjord P, Kalamajski S. The novel small leucine-rich protein chondroadherin-like (CHADL) is expressed in cartilage and modulates chondrocyte differentiation. J Biol Chem. 2015;290(2):918–25.
2. Hessle L, Stordalen GA, Wenglén C, Petzold C, Tanner E, Brorson SH, et al. The skeletal phenotype of chondroadherin deficient mice. PLoS One. 2014;8(6):e63080.
3. Mizuno M, Fujisawa R, Kuboki Y. Bone chondroadherin promotes attachment of osteoblastic cells to solid-state substrates and shows affinity to collagen. Calcif Tissue Int. 1996;59(3):163–7.
4. Johnson JM, Young TL, Rada JA. Small leucine rich repeat proteoglycans (SLRPs) in the human sclera: identification of abundant levels of PRELP. Mol Vis. 2006;12:1057–66.
5. Zhou HY. Proteomic analysis of hydroxyapatite interaction proteins in bone. Ann N Y Acad Sci. 2007;1116:323–6.
6. Paracuellos P, Kalamajski S, Bonna A, Bihan D, Farndale RW, Hohenester E. Structural and functional analysis of two small leucine-rich repeat proteoglycans, fibromodulin and chondroadherin. Matrix Biol. 2017;63:106–16.
7. Camper L, Heinegârd D, Lundgren-Akerlund E. Integrin alpha2beta1 is a receptor for the cartilage matrix protein chondroadherin. J Cell Biol. 1997;138(5):1159–67.
8. Haglund L, Tillgren V, Addis L, Wenglén C, Recklies A, Heinegård D. Identification and characterization of the integrin alpha2beta1 binding motif in chondroadherin mediating cell attachment. J Biol Chem. 2011;286(5):3925–34.
9. Haglund L, Tillgren V, Önnerfjord P, Heinegård D. The C-terminal peptide of chondroadherin modulates cellular activity by selectively binding to heparan sulfate chains. J Biol Chem. 2013;288(2):995–1008.
10. Shen Z, Gantcheva S, Mânsson B, Heinegârd D, Sommarin Y. Chondroadherin expression changes in skeletal development. Biochem J. 1998;330((Pt 1)(Pt 1)):549–57.
11. Batista MA, Nia HT, Önnerfjord P, Cox KA, Ortiz C, Grodzinsky AJ, et al. Nanomechanical phenotype of chondroadherin-null murine articular cartilage. Matrix Biol. 2014;38:84–90.
12. Tasheva ES, Ke A, Conrad GW. Analysis of the expression of chondroadherin in mouse ocular and non-ocular tissues. Mol Vis. 2004;10:544–54.
13. Landgren C, Beier DR, Fässler R, Heinegård D, Sommarin Y. The mouse chondroadherin gene: characterization and chromosomal localization. Genomics. 1998;47(1):84–91.
14. Pesch K, Zeitz C, Fries JE, Münscher S, Pusch CM, Kohler K, et al. Isolation of the mouse nyctalopin gene nyx and expression studies in mouse and rat retina. Invest Ophthalmol Vis Sci. 2003;44(5):2260–6.
15. Bech-Hansen NT, Naylor MJ, Maybaum TA, Sparkes RL, Koop B, Birch DG, et al. Mutations in NYX, encoding the leucine-rich proteoglycan nyctalopin, cause X-linked complete congenital stationary night blindness. Nat Genet. 2000;26(3):319–23.

16. O'Connor E, Eisenhaber B, Dalley J, Wang T, Missen C, Bulleid N, et al. Species specific membrane anchoring of nyctalopin, a small leucine-rich repeat protein. Hum Mol Genet. 2005;14(13):1877–87.
17. Pusch CM, Zeitz C, Brandau O, Pesch K, Achatz H, Feil S, et al. The complete form of X-linked congenital stationary night blindness is caused by mutations in a gene encoding a leucine-rich repeat protein. Nat Genet. 2000;26(3):324–7.
18. Miyake Y, Yagasaki K, Horiguchi M, Kawase Y, Kanda T. Congenital stationary night blindness with negative electroretinogram. A new classification. Arch Ophthalmol. 1986;104(7):1013–20.
19. Kofuji P, Ceelen P, Zahs KR, Surbeck LW, Lester HA, Newman EA. Genetic inactivation of an inwardly rectifying potassium channel (Kir4.1 subunit) in mice: phenotypic impact in retina. J Neurosci. 2000;20(15):5733–40.
20. Gregg RG, Mukhopadhyay S, Candille SI, Ball SL, Pardue MT, McCall MA, et al. Identification of the gene and the mutation responsible for the mouse nob phenotype. Invest Ophthalmol Vis Sci. 2003;44(1):378–84.
21. Masu M, Iwakabe H, Tagawa Y, Miyoshi T, Yamashita M, Fukuda Y, et al. Specific deficit of the ON response in visual transmission by targeted disruption of the mGluR6 gene. Cell. 1995;80(5):757–65.
22. Cao Y, Posokhova E, Martemyanov KA. TRPM1 forms complexes with nyctalopin in vivo and accumulates in postsynaptic compartment of ON-bipolar neurons in mGluR6-dependent manner. J Neurosci. 2011;31(32):11521–6.
23. Istiaq A, Ohta K. A review on Tsukushi: mammalian development, disorders, and therapy. J Cell Commun Signal. 2022;16(4):505–13.
24. Matsushima N, Miyashita H, Kretsinger RH. Sequence features, structure, ligand interaction, and diseases in small leucine rich repeat proteoglycans. J Cell Commun Signal. 2021;15(4):519–31.
25. Ahmad SAI, Anam MB, Ito N, Ohta K. Involvement of Tsukushi in diverse developmental processes. J Cell Commun Signal. 2018;12(1):205–10.
26. Miwa T, Ohta K, Ito N, Hattori S, Miyakawa T, Takeo T, et al. Tsukushi is essential for the development of the inner ear. Mol Brain. 2020;13(1):29.
27. Niimori D, Kawano R, Felemban A, Niimori-Kita K, Tanaka H, Ihn H, et al. Tsukushi controls the hair cycle by regulating TGF-β1 signaling. Dev Biol. 2012;372(1):81–7.
28. Yano K, Washio K, Tsumanuma Y, Yamato M, Ohta K, Okano T, et al. The role of Tsukushi (TSK), a small leucine-rich repeat proteoglycan, in bone growth. Regen Ther. 2017;7:98–107.
29. Hattori T, Müller C, Gebhard S, Bauer E, Pausch F, Schlund B, et al. SOX9 is a major negative regulator of cartilage vascularization, bone marrow formation and endochondral ossification. Development. 2010;137(6):901–11.
30. Bi Y, Stuelten CH, Kilts T, Wadhwa S, Iozzo RV, Robey PG, et al. Extracellular matrix proteoglycans control the fate of bone marrow stromal cells. J Biol Chem. 2005;280(34):30481–9.
31. Zamurovic N, Cappellen D, Rohner D, Susa M. Coordinated activation of notch, Wnt, and transforming growth factor-beta signaling pathways in bone morphogenic protein 2-induced osteogenesis. Notch target gene Hey1 inhibits mineralization and Runx2 transcriptional activity. J Biol Chem. 2004;279(36):37704–15.
32. Low SWY, Connor TB, Kassem IS, Costakos DM, Chaurasia SS. Small leucine-rich proteoglycans (SLRPs) in the retina. Int J Mol Sci. 2021;22(14):7293.
33. Ohta K, Ito A, Kuriyama S, Lupo G, Kosaka M, Ohnuma S, et al. Tsukushi functions as a Wnt signaling inhibitor by competing with Wnt2b for binding to transmembrane protein Frizzled4. Proc Natl Acad Sci USA. 2011;108(36):14962–7.
34. Li H, Corrales CE, Wang Z, Zhao Y, Wang Y, Liu H, et al. BMP4 signaling is involved in the generation of inner ear sensory epithelia. BMC Dev Biol. 2005;5:16.

35. Ohyama T, Basch ML, Mishina Y, Lyons KM, Segil N, Groves AK. BMP signaling is necessary for patterning the sensory and nonsensory regions of the developing mammalian cochlea. J Neurosci. 2010;30(45):15044–51.
36. Dabdoub A, Puligilla C, Jones JM, Fritzsch B, Cheah KS, Pevny LH, et al. Sox2 signaling in prosensory domain specification and subsequent hair cell differentiation in the developing cochlea. Proc Natl Acad Sci USA. 2008;105(47):18396–401.

Regulatory Role and Significance of Class V SLRPs: Podocan and Podocan-Like Protein-1

7

Arundhati Bag

Abstract

Class V SLRP has two members, podocan and podocan-like protein-1. They bind with collagens just like the other members of SLRPs and although they have leucine-rich repeat domains, they differ in their C-terminal ends and are also considered to act as glycoproteins. This chapter discusses the interaction of class V SLRPs with different signalling pathways, including Wnt/β-catenin, TGF-β, and PI3K/AKT/mTOR where (P) podocan and podocan-like protein-1 have a regulatory effect, and podocan is a negative regulator of cell proliferation and migration, and it can bring these effects by downregulating Wnt/β-catenin signalling. Involvement of class V SLRPs in various diseases, including diabetic nephropathy and different types of cancers is discussed including (E) epigenetic regulation of podocan and podocan-like protein-1.

Keywords

Small Leucine Rich Proteoglycans · Regulation · Cancer · Epigenetics · Podocytes

---

A. Bag (✉)
Department of Medical Biotechnology, Sikkim Manipal Institute of Medical Sciences, Sikkim Manipal University, Gangtok, Sikkim, India
e-mail: arundhati.b@smims.smu.edu.in

P. Lama (ed.), *The Biological Role of Small Leucine-Rich Proteoglycans (SLRPs) in Matrix Homeostasis and Diseases*,
https://doi.org/10.1007/978-981-96-7890-7_7

7.1 Introduction

As mentioned in Chap. 1, class V small leucine-rich proteoglycans (SLRP) have two known members, podocan and podocan-like protein-1. Class V proteins are more recently discovered members of SLRPs. Podocan was discovered in 2003 in podocytes of HIV-1-associated nephropathy, while podocan-like protein-1 was discovered in 2011 in mineralized tissues and osteoblasts. Since then, they have been studied for their role in various cellular functions and disease conditions. They bind to the collagen just like other SLRPs; however, they are categorized as noncanonical proteins having unique terminal caps. They have a higher number of leucine-rich repeat (LRR) domains in comparison to the other classes of SLRPs. Human podocan is known to have twenty LRR domains, and podocan-like protein-1 have twenty one, while the number varies from ten to fifteen in members of other classes of SLRPs. Both of these proteins play an instrumental role in matrix assembly and modulates key signalling pathways in diseases of the matrix.

7.2 Podocan

Human podocan is extracellularly secreted SLRP, consisting of 613 amino acids. Its gene (*PODN*) is located at 1p32.3 and has 13 exons. Podocan has a unique amino-terminal cysteine-rich cluster (CX3CXCX7C), followed by LRR domain. Unlike other SLRPs, it lacks carboxy-terminal cysteine residues, and has highly charged acidic C-terminal, which has potential role in selection in glomerular filtration barrier [1], where it was demonstrated that podocan is a glycoprotein with N- linked oligosaccharides rather than a classical proteoglycan. They bind with collagen I, the most common collagen type making up to 90% proteins of human body. Other than in podocytes, podocan is also expressed in vascular endothelial cells, and vascular smooth muscle cells [2]. Shimizu-Hirota et al., demonstrated with RT-PCR analysis that podocan was weakly but consistently expressed in kidney while it was more robustly expressed in heart, liver and pancreas [3]. Podocan may partly contribute in the fibrillogenesis in the mature glomerular basement membrane and thus contribute to maintaining the integrity of glomerular filtration barrier [4]. Podocan is also highly expressed in mice and human adipose tissue, and is regulated by a cytokine tumor necrosis factor α (TNF α) in mouse adipocytes [5]. Podocan was also detected in mouse plasma which was found to increase in diet-induced obese mice and decrease in obese-diabetic mice. Like other SLRPs, podocan can regulate cell proliferation and migration. Cell migration is crucial in many biological functions including embryonic development, repairing injured tissues, blood vessels formations, immune response activation where podocan is a negative regulator of cell migration and proliferation. It has been found that silencing podocan gene expression could promote cell division by altering cyclin dependent kinase-2 (Cdk 2) and p21 pathways [6].

7.2.1 Cell Signalling Through Wnt/β-Catenin Pathway by Podocan

The Wnt/β-catenin signaling pathway plays an important role in embryonic development and maintaining homeostasis in adult tissue [7]. Aberration in this pathway can lead to various diseases. Wingless and Int-related protein (Wnt) signalling follows canonical and noncanonical paths. The canonical path or Wnt/β-catenin signalling takes place by binding of cysteine-rich Wnt glycoprotein to cell surface receptor, transportation of β-catenin to the nucleus, and activation of T-cell factor/lymphoid enhancer factor (TCF/LEF) to regulate cell proliferation. The noncanonical pathway is independent of β-catenin TCF/LEF and regulates cell polarity and migration, and these two paths are mutually regulated. Thus, Wnt pathways are highly conserved and establish crosstalk with other signalling factors [8].

Podocan was found to act as a negative regulator of smooth muscle cell proliferation and migration in both mice and human cells [9]. Proliferation and migration of smooth muscle cells are critical factors in repairing of vascular wounds. Excessive arterial repair and prolonged smooth muscle cell proliferation were observed in human atherosclerosis in the absence of podocan. This was mediated by activation of the Wnt/β-catenin pathway with nuclear transport of β-catenin [10]. Podocan overexpression downregulates the Wnt/β-catenin path, which significantly reduce the proliferation and migration of human smooth muscle cells. It has been suggested that podocan is a negative regulator or inhibitor of Wnt- pathway.

Podocan can also play important role in the differentiation of bovine skeletal muscle satellite cells by regulating the Wnt/β-catenin pathway [11] and in skeletal muscle where podocan controls cell growth and regeneration. Muscle- derived satellite cells are instrumental in growth as well as repairing of skeletal muscles in adults. These satellite cells are unipotent stem cell and precursor of skeletal muscle tissue. It was found that podocan was upregulated significantly during differentiation of bovine muscle-derived satellite cells [12] Podocan overexpression is thus directly related to the differentiation of the cells, and its inhibition suppresses this activity [13]. So far 19 Wnt members have been identified in humans and (A) Wnt4 is a highly secretory member that can acts as a ligand in wound healing and angiogenesis [14].

Wnt-signalling also plays a role in arterial fibrillation and (P) progression to chronic arterial fibrillation that associates with high levels of podocan and dickkopf-1 (DKK-1), an inhibitor of Wnt-signalling. High level of podocan and DKK-1 are also found to be significantly associated with the increased risk of major adverse cardiac events [15].

7.2.2 TGF-β Signalling Pathway

The transforming growth factor-β (TGF-β) cytokines constitute a conserved family of signalling molecules that play an important role in the development and homeostasis of most adult tissues. Literature indicates that TGF-β has an intriguing role in skeletal muscle tissue regeneration where it is known to inhibit muscle regeneration and promote fibrosis; however, in some studies, it was found that inhibiting TGF-β led to incomplete muscle regeneration [16]. Podocan leads to myogenic differentiation *in vitro* by inhibiting the TGF-β pathway [17]. Podocan was also found to promote post-injury regeneration of skeletal muscle in mice in the same study. The mechanism most likely involves the regulation of two intracellular signaling components TGF-β pathway, p-Smad2 and p-Smad4, and their interaction with TGF-β1.

7.3 Podocan-Like Protein-1

Podocan-like protein-1 was first discovered by a bioinformatic analysis, and the protein comprise of 512 amino acids. The gene of PODNL-1 has a chromosomal location at 19p13.12 with nine exons. PODNL-1 is considered to act as a glycoprotein and it has a unique cysteine-rich N-terminal cluster, with twenty one leucine-rich repeat motifs, and an N-glycosylation site [18]. It is found in bone, tibial nerves, coronary arteries, bone marrow and mesenchymal stem cells and associates with insoluble matrix [18, 19]. It is also an important component of tumor microenvironment [20].

7.3.1 Cell Signalling Through TGF-β Pathway by Podocan-Like Protein-1

TGF-β is a multifunctional cytokine expressed in almost every cell type. TGF-β signalling is at the center of almost every cellular function and thus is crucial in health. TGF-β has been found to be particularly important in embryonic development, tissue homeostasis, regeneration, and in maintaining homeostasis of the immune system [21]. Therefore, its aberrant action can lead to a wide array of diseases, including different types of cancers. It is involved in inflammation, epithelial to mesenchyme transition, angiogenesis, metastasis, and fibroblast activation in the tumor microenvironment. Inactive TGF-β is present in the extracellular matrix and, when activated, binds to its cell surface receptors. Like Wnt-signalling, TGF-β also follows canonical and noncanonical pathways. The canonical pathway recruits Smad2 and/ or Smad3 and transports them to the nucleus to regulate target gene transcription, whereas the noncanonical pathways recruit ERK/MAP kinase to regulate gene expression [22].

TGF-β has a paradoxical role in tumors; where it acts as a tumor suppressor in the early stages but switches to tumor promoter activity in advanced stages of cancer and promotes cancer growth, metastasis, recurrence, and drug resistance [23]. It is found that switching of TGF-β signalling from the canonical to noncanonical pathway can trigger tumor development. It has also been demonstrated that elevated expression of PODNL-1 significantly correlatess with downregulation of the canonical TGF-β pathway, thus promoting tumor progression [24].

7.3.2 Akt/mTOR Pathway

The PI3K/AKT/mTOR is a highly conserved signal transduction pathway that promotes cell cycle, cell growth, and survival. Both AKT and mTOR are serine/threonine kinases and mTOR has two complexes, mTORC1 and mTORC2. *PODNL1* promotes aggressiveness in glioma by activating the Akt/mTOR pathway [25] where mTORC1 downstream PI3kinase and AKT induces glioma cell growth, and mTORC2 also promotes cell proliferation and survival through the regulation of the Akt pathway [26]. It has also been demonstrated that knockdown of PODNL1 inactivates the Akt/mTOR signaling pathway [27].

7.4 SLRP Class V Proteins in Health and Disease

7.4.1 In Cancer

Since SLRPs play an important role in ECM assembly, collagen fibrillogenesis, and signaling for matrix homeostasis, they have been much discussed in scientific literature for their role in tumor development and progression. Studies show that podocan is a potential prognostic marker of malignancy [28, 29]. Right- and left-sided colon cancer vary in their pathology, genetics, epidemiology, severity, as well as in prognosis [30]. SLRP transcript levels were studied in nonneoplastic and tumor tissues of adenocarcinoma [31], and variable expression of these two SLRPs V proteins was found in normal and tumor tissues of the two sides. Higher level of transcripts of the podocan gene (*PODN*) was found in normal tissues of the left colon than of the right colon. Podocan was significantly downregulated in left-sided carcinoma and it is the only noncanonical SLRP whose downregulation was associated with relapse-free survival. Podocan-like protein-1 gene (*PODNL1*) transcripts were not detected in normal as well as in tumor tissue from the right colon. However, PODNL1 transcripts were found in 50% of left-sided colon cancer tissues. In head and neck squamous cell carcinoma (HNSCC), cancer stem cells are located in the invasive fronts close to blood vessels. Research shows that these stem cells are major factors in influencing the recurrence and progression of HNSCC. In bioinformatic analysis, it was shown that *PODN* is one of the twenty key genes related to stemness. In another bioinformatic analysis related to the prognosis of muscle-invasive bladder cancer with patients receiving immunotherapy, *podocan* expressions was found to be oncogenic [32, 33]. In bioinformatics analysis, podocan was also found to be a potential diagnostic as well as prognostic biomarker in osteosarcoma [34] and their expressions correlate with immune infiltrates in bone tissues. In osteosarcoma, podocan was hypermethylated and expressed in a low level, and overexpressions prevented proliferation, migration, and invasion of osteosarcoma cells. Podocan overexpression inactivates the TGF-β/Smad2/3 pathway and suppresses the development of the disease both *in vitro* and *in vivo* in mice. Gastric cancer is the third major cause of cancer-related death worldwide [35] gene was found to be hypermethylated and downregulated in gastric cancer and in silico study was associated with overall survival [2]. Increased downregulation of podocan was associated with better prognosis in gastric cancer, and the podocan gene is a potent regulator of cellular phenotype in the extracellular matrix. Other than methylation, the PODN gene was also affected by amplification, deletion and mutations in gastric cancer. The gene was also downregulated in pancreatic tumor tissue as well as in cell lines [17] and it has been dentified as a tumor suppressor gene that downregulates in tumors, and overexpression was associated with prolonged overall survival.

Bioinformatic analysis for clinical outcome found that upregulation of podocan, along with two other hub proteins (interact with at least ten other proteins) and brought good survival in breast cancer [28]. Whereas downregulation of podocan was an indicator of good survival in ovarian cancer, gastric cancer, and lung cancer.

Podocan-like protein-1 has been more commonly associated with glioma, a malignant brain tumor of glial cells. Glioma is grouped into low-grade glioma or high-grade glioma. PODNL1 was found to be highly expressed in high-grade glioma [36]. Podocan-like protein-1 is a promising predictive indicator of poor prognosis in glioma patients [15–17]. Knockdown and silencing of PODNL1 promotes apoptosis and inhibits cell proliferation and migration. Its overexpression increase mesenchymal markers like N-cadherin, vimentin, snail, and fibronectin. Thus, it promotes cell proliferation, migration, and epithelial to mesenchymal transition in glioma cells. PODNL1 induce aggressive glioma by activating the Akt/mTOR signalling pathway. The effect of silencing PODNL1 on cell proliferation and migration was partially resumed by applying Akt-activator [18] and recent (G) evidence shows that the Akt/mTOR signalling pathway directly regulates malignant behaviors in glioma [12]. Extensive bioinformatic analysis performed to identify drug targets in glioblastoma multiforme, a high-grade glioma, shows that PODNL1, along with twenty other genes, were highly expressed in this glioblastoma multiforme [24]. They were also significantly associated with the low survivability of these patients, and especially elevated PODNL-1 was strongly associated with poor overall survival. Furthermore, bioinformatic analysis on PODNL-1 methylation in lower-grade glioma showed that low methylation, thus high PODNL1 expression was associated with unfavorable prognosis and tumor aggressiveness (Noor 2021). Four CpG sites, CpG sites, cg07425555, cg26969888, cg18547299, and cg24354933, were found to be associated with unfavorable overall survival and disease-free survival. Low PODNL-1 methylation was potentially associated with high immune cell infiltration and inflammatory response in glioma, which may lead to shorter survival in patients with high immune infiltration [10]. High PODNL1 expression due to low methylation in CpG sites was significantly associated with an increase in PD-L1, PD-1, and CTLA4 immune checkpoint genes. It is found that, unlike other types of cancers, lymphocyte infiltration in the tumor site is a negative prognostic factor in glioma [20, 21]. Thus, it is important to understand the immune microenvironment of the tumor before immunotherapeutic intervention.

Recent evidence indicate extracellular matrix proteins have been discussed mainly in the context of tumor metastasis, progression, and immune cell infiltration in tumor tissues. These proteins are not only responsible for maintaining architectural or biochemical conditions of the tumor microenvironment but also can act as cytokines, controlling signal transduction in the tumour microenvironment. Therefore, they have received considerable attention in recent days as markers of immunotherapy outcomes. In a pan-cancer setting, the podocan-like protein-1 was studied by bioinformatic tools for prognostic value and role in tumor microenvironment [17]. Overall, it was found as a marker for poor prognosis and immunotherapeutic responses in various cancers. Its overexpression was associated with poor survival in patients treated with immune-checkpoint blockade-based immunotherapies commonly targeting PD-1, PD-L1, and CTLA-4. PODNL-1 expression was found to be positively correlate with cancer-associated fibroblast (CAF) infiltration levels in thirty-three types of cancers. CAFs are a key cellular component of tumor microenvironment and are responsible for matrix remodeling in tumor, interacting with cancer cells and establishing cross talks with infiltrating immune cells like leukocytes. High PODNL-1 was also associated with infiltration of hematopoietic stem cells, macrophages, and myeloid-derived suppressor cells in many cancer types. Myeloid-derived suppressor cells suppress T-cell activities in tumor microenvironment and thus cause overall immune suppression. High PODNL-1 was also positively associated with expression of TGFβ, TNFα, inflammatory response, hedgehog, and EMT. Further, PODNL-1 was highly expressed in fifteen types of cancers, including breast, head and neck, esophageal, gastric, colorectal, pancreatic, lung, sarcoma, etc., while it was significantly downregulated in nine cancer types [21, 35]. PODNL-1 was also found to be potentially involved in collagen dysfunction in ECM remodeling in the study.

7.4.2 In Kidney Disease

Podocan is expressed in kidney glomeruli and maintains the integrity of the normal glomerular filtration barrier, and may cause fibrillogenesis of collagens in the mature glomerular basement membrane [7, 8]. Podocan mRNA is visible in the cytoplasm of podocytes, and the outermost epithelial cell layer of the glomerular filtration barrier. It was also visible in vascular endothelial cells within the glomerulus [10]. The glomerular filtration barrier is constitute of the outermost epithelial layer of podocytes, followed by the glomerular basement membrane and innermost endothelial cell layer [13].

Diabetic nephropathy, a progressive kidney disease in diabetic conditions, is a major cause of kidney failure worldwide. One of the remarkable changes in diabetic nephropathy is glomerular hyperfiltration, which leaks albumin into urine at a greater-than-normal level and is also known as albuminuria. Other phenotypes of diabetic nephropathy include glomerular hypertrophy, mesangial expansion, thickening of the basement membrane, and tubular hyperfiltration. One of the major causes of albuminuria is the disintegration of glomerular filtration barrier in a high-glucose environment. Since podocan is present in epithelial and endothelial component layers and interacts with the collagen in the matrix, any changes in its expression would also alter the structure of the filtration barrier. In the diabetic nephropathy mouse model, podocan mRNA was found to be decreased in the renal cortex as well as in plasma. A decrease in podocan is increase albuminuria, which indicates a damaged glomerular filtration barrier and marks diabetic nephropathy.

Podocan has anti-fibrotic activity in the glomerulus and TGF-β elevated podocan gene expression was reported in an *in vitro* human mesangial cells [27]. The podocan, in turn, downregulates the expression of profibrotic genes such as collagen 1A1, fibronectin, and connective tissue growth factor CTGF and has a mild negative regulation on fibrosis. However, it didn't cause fibrinolysis in the mesangial cells.

7.4.3 Other Diseases

Placenta accreta is a pathologic condition in which chorionic villi adhere deeply to or invade the uterine wall. It may cause maternal morbidity and mortality. Podocan, a negative regulator of cell migration, was studied for placenta accreta [22]. Podocan expression was downregulated at both transcriptome and proteome levels, which induced trophoblast migration, causing placenta accreta. Podocan was also detected in the aortic intima, expressed in cardiac and vascular smooth muscle cells. As mentioned earlier, a lack of podocan results in excessive arterial repair following injury and prolonged smooth muscle cell proliferation [37]. Podocan was studied with risk of adverse outcomes in chest pain with suspected acute coronary syndromes, as podocan has a role in the fibrosis of blood vessels. However, podocan was not associated with all-cause mortality or other major cardiovascular adverse outcomes in patients with acute chest pain [31].

7.5 Epigenetic Regulation of Podocan and Podocan-Like Protein-1

While podocan and podocan-like protein-1 regulate different signalling pathways, they themselves are regulated epigenetically. In the past few years, it has been increasingly understood that epigenetic regulation is an integral aspect of gene expression, and its aberrant functioning can lead to various metabolic diseases. Epigenetic regulations can take place by DNA methylation, histone modification, and noncoding RNA. MicroRNA (miRNA) is a small (approximately 22 nucleotides) non-coding RNA that modifies genes posttranscriptionally. More than 2000 miRNAs have been identified in man [1] and it was found that microRNA, miR-3180-5p, was overexpressed in human bladder smooth muscle cells under hydrodynamic pressure [38]. Human bladder smooth muscle cells are the major cell type in the bladder, which sense the hydrostatic pressure on the bladder. Overexpressed miR-3180-5p downregulates the expression of *podocan and* podocan like protein-1, resulting in cell proliferation by upregulation of cyclin-dependent kinase-2 and downregulation of cell cycle inhibitor p21 [29]. miR-1908 negatively regulates podocan-like protein-1 gene. High expression of miR-1908 and low expression of podocan like protein-1 are associated with longer survival in ovarian cancer, as podocan-like protein-1 has a potential role in tumor development and progression [31]. The *PODNL-1* gene was also found to be regulated by miR196d in crucian carp model [37, 38]. Thus, as discussed above, methylation and expression of both *PODN* and *PODNL-1* genes are altered in various disease conditions.

References

1. Alles J, Fehlmann T, Fischer U, Backes C, Galata V, Minet M, Hart M, Abu-Halima M, Grässer FA, Lenhof HP, Keller A, Meese E. An estimate of the total number of true human miRNAs. Nucleic Acids Res. 2019;47(7):3353–64. https://doi.org/10.1093/nar/gkz097.
2. Andersen T, Ueland T, Aukrust P, Nilsen DW, Grundt H, Staines H, Kontny F. Podocan and adverse clinical outcome in patients admitted with suspected acute coronary syndromes. Front Cardiovasc Med. 2022;9:867944. https://doi.org/10.3389/fcvm.2022.867944.
3. Bai Y, Wei C, Zhong Y, Zhang Y, Long J, Huang S, Xie F, Tian Y, Wang X, Zhao H. Development and palidation of a prognostic nomogram for gastric cancer based on DNA methylation-driven differentially expressed genes. Int J Biol Sci. 2020;16(7):1153–65. https://doi.org/10.7150/ijbs.41587.
4. Baran B, Mert Ozupek N, Yerli Tetik N, Acar E, Bekcioglu O, Baskin Y. Difference between left-sided and right-sided colorectal cancer: a focused review of literature. Gastroenterology Res. 2018;11(4):264–73.
5. Shrestha B, Watthanasuntorn K, Bathini TJ, Gopal S, Panchal A, Patel H, Gilani A, Victory J, Scribani M, Hutter R, Katz DH. Association between Wnt-pathway molecules (podocin, Dkk1 and CRP) and major adverse cardiac events- 5year follow up analysis. J Am Coll Cardiol. 2024;83:84.
6. Teng C, Zheng H. Low expression of microRNA-1908 predicts a poor prognosis for patients with ovarian cancer. Oncol Lett. 2017;14:4277–81.
7. Clayton SW, Ban GI, Liu C, Serra R. Canonical and noncanonical TGF-β signaling regulate fibrous tissue differentiation in the axial skeleton. Sci Rep. 2020;10(1):21364. https://doi.org/10.1038/s41598-020-78206-4.
8. Dai G, Sun Y, Wei R, Xi L. Small leucine-rich proteoglycan *PODNL1* identified as a potential tumor matrix-mediated biomarker for prognosis and immunotherapy in a pan-cancer setting. Curr Issues Mol Biol. 2023;45(7):6116–39. https://doi.org/10.3390/cimb45070386.
9. Deng Z, Fan T, Xiao C, Tian H, Zheng Y, Li C, He J. TGF-β signaling in health, disease, and therapeutics. Signal Transduct Target Ther. 2024;9(1):61. https://doi.org/10.1038/s41392-024-01764-w.
10. Didangelos A, Yin X, Mandal K, Baumert M, Jahangiri M, Mayr M. Proteomics characterization of extracellular space components in the human aorta. Mol Cell Proteomics. 2010;9(9):2048–62. https://doi.org/10.1074/mcp.M110.001693.
11. Geng Y, Zuo P, Li XO, Zhang L. PODNL1 promotes cell proliferation and migration in glioma via regulating Akt/mTOR pathway. J Cancer. 2020;11(21):6234–42. https://doi.org/10.7150/jca.46901.
12. Han X, Zhang Y, Lu F, Feng J, Zhang C, Wang G. Hypermethylated PODN represses the progression of osteosarcoma by inactivating the TGF-β/Smad2/3 pathway. Pathol Res Pract. 2022;238:154075. https://doi.org/10.1016/j.prp.2022.

13. Hutter R, Huang L, Speidl WS, Giannarelli C, Trubin P, Bauriedel G, Klotman ME, Fuster V, Badimon JJ, Klotman PE. Novel small leucine-rich repeat protein podocan is a negative regulator of migration and proliferation of smooth muscle cells, modulates neointima formation, and is expressed in human atheroma. Circulation. 2013;128(22):2351–63. https://doi.org/10.1161/CIRCULATIONAHA.113.004634.
14. Kim J, Lee J. Role of transforming growth factor-β in muscle damage and regeneration: focused on eccentric muscle contraction. J Exerc Rehabil. 2017;13(6):621–6. https://doi.org/10.12965/jer.1735072.536.
15. Li M, Cheng J, Ma Y, Guo H, Shu H, Huang H, Kuang Y, Yang T. The histone demethylase JMJD2A promotes glioma cell growth via targeting Akt-mTOR signaling. Cancer Cell Int. 2020;20:101. https://doi.org/10.1186/s12935-020-01177-z.
16. Li N, Hou R, Liu C, Yang T, Qiao C, Wei J. Integration of transcriptome and proteome profiles in placenta accreta reveals trophoblast over-migration as the underlying pathogenesis. Clin Proteomics. 2021;18(1):31. https://doi.org/10.1186/s12014-021-09336-8.
17. Li S, Liu D, Fu Y, Zhang C, Tong H, Li S, Yan Y. Podocan promotes differentiation of bovine skeletal muscle satellite cells by regulating the Wnt4-β-catenin signaling pathway. Front Physiol. 2019;10:1010. https://doi.org/10.3389/fphys.2019.01010.
18. Yin G. Wnt/β-catenin signalling: function, biological mechanisms, and therapeutic opportunities. Signal Transduct Target Ther. 2022;7(1):3. https://doi.org/10.1038/s41392-021-00762-6.
19. Lyskjær I, Kara N, De Noon S, Davies C, Rocha AM, Strobl AC, Usher I, Gerrand C, Strauss SJ, Schrimpf D, von Deimling A, Beck S, Flanagan AM. Osteosarcoma: novel prognostic biomarkers using circulating and cell-free tumour DNA. Eur J Cancer. 2022;168:1–11. https://doi.org/10.1016/j.ejca.2022.03.002.
20. Marinari E, Allard M, Gustave R, Widmer V, Philippin G, Merkler D, Tsantoulis P, Dutoit V, Dietrich PY. Inflammation and lymphocyte infiltration are associated with shorter survival in patients with high-grade glioma. Onco Targets Ther. 2020;9(1):1779990. https://doi.org/10.1080/2162402X.2020.1779990.
21. Mochida Y, Kaku M, Yoshida K, Katafuchi M, Atsawasuwan P, Yamauchi M. Podocan-like protein: a novel small leucine-rich repeat matrix protein in bone. Biochem Biophys Res Commun. 2011;410(2):333–8. https://doi.org/10.1016/j.bbrc.2011.05.150.
22. Nio Y, Okawara M, Okuda S, Matsuo T, Furuyama N. Podocan is expressed in blood and adipose tissue and correlates negatively with the induction of diabetic nephropathy. J Endocr Soc. 2017;1(7):772–86. https://doi.org/10.1210/js.2017-00123.
23. Ross MD, Bruggeman LA, Hanss B, Sunamoto M, Marras D, Klotman ME, Klotman PE. Podocan, a novel small leucine-rich repeat protein expressed in the sclerotic glomerular lesion of experimental HIV-associated nephropathy. J Biol Chem. 2003;278(35):33248–55. https://doi.org/10.1074/jbc.M301299200.
24. Shergalis A, Bankhead A 3rd, Luesakul U, Muangsin N, Neamati N. Current challenges and opportunities in treating glioblastoma. Pharmacol Rev. 2018;70(3):412–45. https://doi.org/10.1124/pr.117.014944.
25. Shi X, Yang J, Deng S, Xu H, Wu D, Zeng Q, Wang S, Hu T, Wu F, Zhou H. TGF-β signaling in the tumor metabolic microenvironment and targeted therapies. J Hematol Oncol. 2022;15(1):135. https://doi.org/10.1186/s13045-022-01349-6.
26. Shimizu-Hirota R, Sasamura H, Kuroda M, Kobayashi E, Saruta T. Functional characterization of podocan, a member of a new class in the small leucine-rich repeat protein family. FEBS Lett. 2004;563(1–3):69–74. https://doi.org/10.1016/S0014-5793(04)00250-9.
27. Solis-Hernandez MP, Martín C, García B, Pérez-López N, García-Mesa Y, González- Fernández S, García-Suárez O, Merayo J, Fernández-Vega I, Quirós LM. The genes encoding small leucine-rich proteoglycans undergo differential expression alterations in colorectal cancer, depending on tumor location. Cells. 2021;10(8):2002. https://doi.org/10.3390/cells10082002.
28. Sun Y, Luo DY, Zhu YC, Zhou L, Yang TX, Tang C, Shen H, Wang KJ. MiR 3180-5p promotes proliferation in human bladder smooth muscle cell by targeting PODN under hydrodynamic pressure. Sci Rep. 2016;6:33042. https://doi.org/10.1038/srep33042.

29. Tang R, Liu X, Wang W, Hua J, Xu J, Liang C, Meng Q, Liu J, Zhang B, Yu X, Shi S. Identification of the roles of a stemness index based on mRNA expression in the prognosis and metabolic reprograming of pancreatic ductal adenocarcinoma. Front Oncol. 2021;11:643465. https://doi.org/10.3389/fonc.2021.643465.
30. Teng H, Zheng J, Liang Y, Zhao J, Yan Y, Li S, Li S, Tong H. Podocan promoting skeletal muscle post-injury regeneration by inhibiting TGF-β signaling pathway. FASEB J. 2024;38(5):e23502. https://doi.org/10.1096/fj.202302158RR.
31. Wang Z, Li X, Wang X, Liu J, Wang L, Wei W, Duan X, Ding D. Classification of muscle invasive bladder cancer to predict prognosis of patients treated with immunotherapy. J Immunol Res. 2022;2022:6737241. https://doi.org/10.1155/2022/6737241. PMID: 35677536
32. Wu ZH, Li C, Zhang YJ, Zhou W. Identification of a cancer stem cells signature of head and neck squamous cell carcinoma. Front Genet. 2022;13:814777. https://doi.org/10.3389/fgene.2022.814777.
33. Zhang X, Luo M, Jiang B, Zhu W, Min Q, Hu J, Liu T, Fu J, Shi X, Wang P, Wang L, Dong Z. microRNA regulation of skin pigmentation in golden-back mutant of crucian carp from a rice- fish integrated farming system. BMC Genomics. 2023;24(1):70. https://doi.org/10.1186/s12864-023-09168-w.
34. Yan Y, Zhang L, Xu T, Zhou J, Qin R, Chen C, Zou Y, Fu D, Hu G, Chen J, Lu Y. SAMSN1 is highly expressed and associated with a poor survival in glioblastoma multiforme. PLoS One. 2013;8(11):e81905. https://doi.org/10.1371/journal.pone.0081905.
35. Yang D, Li Q, Shang R, Yao L, Wu L, Zhang M, Zhang L, Xu M, Lu Z, Zhou J, Huang L, Huang X, Cheng D, Yang Y, Yu H. WNT4 secreted by tumor tissues promotes tumor progression in colorectal cancer by activation of the WNT/β-catenin signalling pathway. J Exp Clin Cancer Res. 2020;39(1):251. https://doi.org/10.1186/s13046-020-01774-w.
36. Yao F, Zhu ZF, Wen J, Zhang FY, Zhang Z, Zhu LQ, Su GH, Yuan QW, Zhen YF, Wang XD. PODN is a prognostic biomarker and correlated with immune infiltrates in osteosarcoma. Cancer Cell Int. 2021;21(1):381. https://doi.org/10.1186/s12935-021-02086-5.
37. Zhang F, Wang L, Chen Q, Zhang F, Wang X, Yao F. Podocan unraveled: understanding its role in tumor proliferation and smooth muscle regulation. Biomed Pharmacother. 2024;179:117416. https://doi.org/10.1016/j.biopha.2024.117416.
38. Zhao F, Yu YQ. The prognostic roles of mRNAs of the exosomes derived from bone marrow stromal cells in common malignancies: a bioinformatic study. Onco Targets Ther. 2018;11:7979–86. https://doi.org/10.2147/OTT.S172414.

Biological Basis Behind Molecular Fragmentation of SLRPs as DAMPs

8

Polly Lama and Karma Lakhi Bhutia

Abstract

The small leucine-rich proteoglycans (SLRPs) are extensively localized extracellular matrix molecules that are dynamically synthesized, secreted, deposited, and degraded in vivo. SLRPs are pericellular in location that allows them to interact with other matrix molecules such as collagens, fibronectin, and fibrillin, as well as with various types of cellular receptors and signalling molecules. Nearly all classes of SLRPs regulate the vital process of collagen fibrillogenesis thereby contributing to the maintenance of tissue structure. During injury and inflammation in various tissue types, SLRPs degrade and lose their GAG side chain, and this process also affects aggrecan, a large proteoglycan molecules trapped within the fibrous network of collagen, ultimately altering the biochemical and mechanical structure of the matrix. SLRPs also act as a link molecule in the matrix, and thorough the combined action of increased mechanical loading, injury, and proteolytic activity of the matrix-degrading enzymes, SLRPs become fragmented into smaller sizes, which represents a common end stage of the pathological tissues. These cleaved SLRPs fragments can activate Toll-like receptors to initiate a sterile inflammatory reaction even in the absence of blood vessels, leading to upregulated expression of catabolic mediators and associated signaling pathways. Thus, small leucine-rich proteoglycans can act as damage-associated molecular patterns, as they cluster different types of receptors modulating altered signaling events that links the innate and adaptive immune response, and such changes in SLRPs expression leads to diverse biological consequences within the matrix. Therefore, the aim of this chapter is to summarize the multifactorial evidence behind the role of SLRPs as DAMPs.

P. Lama (✉) · K. L. Bhutia
Department of Anatomy, Sikkim Manipal Institute of Medical Sciences, Sikkim Manipal University, Gangtok, Sikkim, India
e-mail: polly.lama@smims.smu.edu.in

P. Lama (ed.), *The Biological Role of Small Leucine-Rich Proteoglycans (SLRPs) in Matrix Homeostasis and Diseases*,
https://doi.org/10.1007/978-981-96-7890-7_8

Keywords

Small Leucine Rich Proteoglycans · Disease · DAMPs · Glycosaminoglycans · Matrix Metalloproteinases · Peptides

8.1 Introduction to DAMPs

Damage-associated molecular patterns (DAMPs) are endogenous, chemotactic, and immune-activating proteins/peptides that are released during cell injury or death, tissue injury, and as a response to pain and inflammation [1]. DAMPs function as intercellular signals that interact with pattern recognition receptors (PRRs) to recruit immune cells and signaling molecules in host defense. The release of DAMPs can lead to potentially lethal cytokine storm in certain pathologies as they can amplify the mechanisms and action of exogenous pathogen-associated molecular patterns (PAMPs), such as the bacterial lipopolysaccharide (LPS), CpG oligodeoxynucleotide, and flagellin, including deleterious endogenous products of damaged and dead cells, accumulating fragmented matrix molecules of SLRPs, that interact with PRRs, such as the Toll-like receptors (TLRs) (Fig. 8.1) to act as cytokine-like mediators accelerating tissue pathogenesis [2].

DAMPs are categorized as molecules that are released to evoke innate inflammatory and adaptive immune responses [3], and many of these molecules dervied from the host tissue are identified as DAMPs, including endogenous moieties, such as the SLRPs fragments, generated due to proteolysis of the core SLRP protein, capable of activating innate/inflammatory and adaptive immune responses [4, 5]. DAMPs are also defined as "alarmins" and currently include chromatin-binding molecules, high-mobility group box-1 (HMGB-1), IL-1α, and IL-33, as well as heat shock proteins (HSPs), S100 proteins, ATPs, and uric acid crystals, which are classified based on their putative actions.

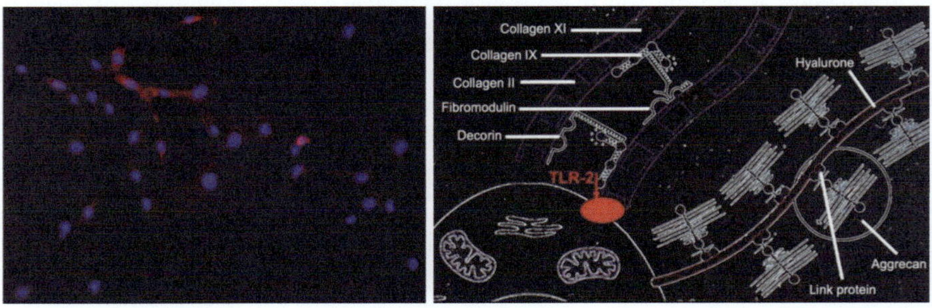

Fig. 8.1 Intervertebral disc cells expressing TLR-2 receptors in cell culture condition (left), blue indicates DAPI-stained cell nuclei and TLR-2 receptors are stained red. Scale bar 100 μm. Schematic representation of TLR-2 receptor interacting with SLRPs and other matrix molecules. Figure adapted from Dr. Polly Lama, Molecular Matrix Lab

8.2 Dysregulation of SLRPs Modifies the Vital Structure of the Matrix

The primary function of SLRPs is to maintain the tissue structure mediated through the process of collagen fibrillogenesis, resulting in delicate corneal transparency, rigidity of osseous and cartilaginous matrix, tensile strength of tendons, and the sponginess of the viscera [6]. In vivo and in vitro analyses show that the arrangement of collagen fibers is disrupted, lost, or increased as the expression of SLRPs changes, which affects the temporospatial matrix organization in cornea, bones, cartilage, muscles, tendon, skin, and various systemic organs during development, traumatic, and nontraumatic pathologies [7]. Further, in rare diseases such as human cornea plana, missense and frameshift mutations occur when the gene coding for SLRP keratocan is not expressed [8], detailed in Chaps. 1, 2, and 6. Similarly, high myopia in ocular tissue is associated with intronic variations of SLRPs genes fibromodulin [9], PRELP [10], and opticin [11]. SLRPs lumican, decorin, and biglycan knockout mice, including compound mutant mice, demonstrate abnormal collagen phenotype with increased numbers of large fibril diameters that are irregular in contours in the cornea [12–15]. Further, in skin, bone, cartilage, and dentine tissue, SLRP-deficient mice exhibit consistent dysfunctional matrix assembly [16–18], including soft tissue organs, such as liver, pancreas, kidney, and the pregnant uterus [19, 20], resembling human diseases and tissue abnormalities. As described in Chaps. 1, 2, and 3, targeted disruption of the biglycan gene leads to osteoporosis-like phenotype in mice [21], and biglycan/fibromodulin-deficient mice show abnormal arrangement of collagen fibrils in tendons that leads to gait impairment, ectopic ossification, and osteoarthritis [22], skin laxity, fragility, as well as joint laxity in decorin and biglycan-deficient mice [23] as well as in lumican and fibromodulin-deficient mice [24]. Further, cancers, osteoporosis, and osteoarthritis, as described in the previous chapters, correspond to altered expression of SLRPs and are also observed in a broad range of other rare human diseases such as Marfan syndrome [25], localized scleroderma [26], infantile progeroid [27], osteogenesis imperfecta [28], and carbohydrate-deficient glycoprotein syndrome [29].

8.2.1 Glycosaminoglycans (GAGs)

Glycosaminoglycans (GAGs) are the essential component of the matrix, and they bind to the SLRPs core protein molecule as side chains interacting with collagen to modulate the process of fibrillogenesis [30, 31]. These linkages are periodic, as the GAG chains that bind to the collagen molecules act as interfibrillar elastic bridges regulating diameter, spacing, and assembly of the collagen fibrils. Previous findings also indicate that enzymatic removal of GAGs may not affect the mechanical properties of collagen fibrils, as GAGs contribute toward absorbing viscoelastic tensile forces in various types of connective tissue [32, 33] except in pathological post-injury tendons [34]. In the mouse model, native and GAG-depleted tendons promote the collagen fibrils sliding under strain by isolating individual fibrils to bear load [35], indicating a variable role of GAG in the extracellular matrix. GAG protect the collagen fibers in acidic conditions against cathepsin K during bone resorption [36], and in articular cartilage, intervertebral disc, skin, and dentine tissue, biglycan, decorin, fibromodulin, and lumican lose their GAG side chain due to increased loading, ageing, and severe tissue degeneration [37, 38]. Further, SLRP osteoadherin exhibits a unique pattern of changes in glycosylation profiles that varies during endochondral bone formation and the biomineralization process [39]. Growth factors and cytokines can also regulate the GAG chain length and sulfation in lumican, and application of interleukin1β (IL-1β) results in a non-glycanated form of lumican in human articular chondrocyte cells, and these findings are comparable to changes as observed with ageing and severe injuries [40]. Thus, changes in length of GAG chains are influenced by basic fibroblast growth factor (bFGF), insulin growth factor (IGF)-1, and transforming growth factor (TGF-β), indicating the complex role of SLRP and its GAG chain in normal and degenerated tissues. In non-mineralized tissue, non-glycanated and N-glycosylated forms of GAGs can increase during growth factor/cytokine synthesis, which is also one of the possible mechanisms explaining the post-translational GAG modifications in the core protein of SLRPs. Non-glycanated SLRPs tend to result from degradation of a glycanated precursor, as demonstrated by the N-terminal sequence analysis of SLRPs in intervertebral disc tissue [37]. These GAG chains are cleaved by matrix-degrading enzymes and serve as a factor for understanding the physiological effect of SLRP residue protein modifications including the molecular level significance of protein-GAG recognition in cell receptor activation during cancer, bone, and cartilage tissue pathogenesis; thus, modification in GAG chain depolymerizes the chemical specificity of the matrix [41].

8.2.2 Matrix metalloproteinases (MMPs)

Matrix metalloproteinases (MMPs) are zinc-dependent enzymes that have been recognized to cleave SLRP members in vitro, and dysregulation of MMPs leads to the development of numerous diseases [42]. MMP-2, MMP-3, and MMP-7 can cleave the human recombinant decorin, and MMP-2 and MMP-3 show decreased efficiency when decorin lacks its GAG chain [43]. Membrane-type matrix metalloproteinase-1 (MT1-MMP) can cleave the human recombinant lumican [44], and similarly fibromodulin is degraded into smaller fragments by MMP-13 in the N-terminal region along with the sulphated tyrosine residue in cartilage tissue disorder, which impacts the stability of the matrix [45]. MMP-13 is also shown to cleave biglycan, opticin, and, to a lesser extent, lumican and decorin, and even though the fragmented peptides are of low abundance, a direct interaction between MMP-13 and decorin suggests that these cleaved peptides can be released to the synovial fluid [46, 47], as described in Chap. 2. In articular cartilage and intervertebral discs, MMP-1, 2, 3, 7, 8, and 9 induces SLRP catabolism with different proteolytic efficiencies [48, 49], with MMP-1, 2, 3, 7, and 13 being the most potent degradative enzymes in cartilage, tendons, and cancers [50–52]. MMP-9 and 12 cleave biglycan in rheumatoid arthritis, and the fragments released are neo-epitopes identified by antibodies and associated with dysregulated matrix remodeling and pathogenesis. Biglycan fragments are also shown to increase ex vivo in degraded cartilage explants and the serum of rats with rheumatoid arthritis [53]. In addition to the MMPs, ADAMTS-4, ADAMTS-5, and granzyme B are the other efficient enzymes that cleave SLRPs biglycan, decorin, and opticin [54–56]. SLRP chondroadherin, additionally, becomes fragmented into smaller sizes by the serine protease HTRA1 in degenerated human intervertebral discs [57]. The nonuniform location of the cleavage sites in different tissue types lead to numerous SLRP fragments, for instance, biglycan is broken down in its N-terminal region [58] of the core protein and the fifth LRR by ADAMTS-4 and ADAMTS-5 [56] and in the C-terminal region by MMP-13 [46]. Therefore, the biological importance of SLRPs, as fragments or degraded products, presents a potential in contributing toward accelerated tissue degradation as these fragments owing to it relatively small molecular size and mass can interact with cellular receptors in diseases, such as cancer and osteoarthritis, initiating downregulation/upregulation of signaling events considered to be detrimental to the tissue structure. These fragments can also accumulate in the matrix and can be detected in serum; thus, SLRP fragments can act as biomarker of various pathological grades and diseases [59–62].

8.3 Fragmented SLRPs and Cell-Matrix Interactions

Altered expressions of SLRPs can convert the cell-matrix phenotypes in various tissue types and diseases, such as cancer [63], osteoarthritis [64], diabetes [65], and even during ageing [66]. Upregulated or downregulated expression of SLRPs compromises the matrix structure, creating a poor biochemical environment with increased expression of cytokines and matrix-degrading enzymes that further contribute toward tissue damage and inflammation, hindering repair and activating signalling pathways, leading to the ingrowths of blood vessels and nerves [67, 68]. Proteolytic fragments of SLRPs biglycan, fibromodulin, decorin, chondroadherin, and lumican bind to the cellular receptors of the host and inflammatory cells, and these fragments, generated as a consequence of inflammatory events, further increase the production of proteases [69]. Matrix-degrading enzymes, such as MMPs 1,2,3,5,7,10, and 13, effectively cleave SLRPs, fibronectin, collagen, and large proteoglycans [70, 71]) to activate Toll-like receptors 2 and 4 [4, 72, 73], resulting in translocation of transcription factor NF kappa B (nuclear factor kappa light chain enhancer of activated B cells) and MAPK (mitogen-activated protein kinases) in many cell types of various tissues that further elicit synthesis of pain mediators, such as prostaglandins, nitric oxide, and phospholipase-A2 capable of inducing changes, even in the peripherally located dorsal root ganglion [74]. Recent research shows that TLR 2 and 4 are particularly activated by soluble biglycan as DAMPs to promote a sterile inflammatory reaction, causing further matrix breakdown and pain [75, 76]. SLRP fragments can thus also be termed as damage-associated molecular patterns (DAMPS) as they act in a similar way as pathogen-associated molecular patterns (PAMPS), triggering TLR inflammatory response without the involvement of leukocytes [5, 77]. SLRP fragments can promote a sterile inflammatory reaction, triggering excessive cytokine and chemokine release and upregulation of associated intercellular transcription factors [78, 79]. Based on the cytoplasmic homology domain, TLRs have been classified to act along with the interleukin receptor (IL-Rs) superfamily, where their role has been defined in terms of pattern recognition molecules identifying the outer layer of pathogens, such as lipopolysaccharide (LPS) on gram-negative bacteria and peptidoglycan (PGN) on gram-positive bacteria [80]. Activation of TLRs on cells involved in immune surveillance triggers both phagocytosis of the pathogenic agent and the release of other cytokines to induce differentiation of the cells involved in the adaptive immune response [73, 81]. Gene expression of TLR 1, 2, 3, 4, 5, 6, 9, and 10 has been identified in cancers [82, 83], articular cartilages [77], and intervertebral disc degeneration [84], and gene expression of TLR 2, 4, and 6 increases with progressive tissue degeneration and pain [85] (Fig. 8.1). Peripheral articular chondrocytes express TLR-2 and 4 and respond to SLRP fragments and LPS-induced degeneration by upregulating factors related to inflammatory catabolic gene expression [80]. Articular cartilage chondrocytes, intervertebral disc cells, and cancer cells modulate TLR 2 and 4 expressions during advanced stages of the diseases, and TLRs can induce pain in conditions such as osteoarthritis and neurological disorders with peripheral nerve injury, traumatic brain injury, cancer, and even stroke. Common to these conditions are increased tissue catabolism, fragmentation of SLRPs, and activation of TLR receptors that accelerate cytokine production, along with chemokines [86, 87] as directly upregulated mediators. Therefore, a plethora of these published findings, with many other significant studies in different tissues and organs, support the hypothesis that TLR modifying drugs could potentially control inflammation and delay further breakdown of matrix, possibly preventing pathogenesis in various tissue/organ-specific conditions/diseases.

8.4 SLRPs as DAMPs

DAMPs as endogenous and heterogenous mediators use more than one receptor to activate their action and some cell receptors can bind to more than one DAMP to sense multi-signaling pathways during inflammation [72]. SLRPs also show multivalent binding capacities, which are required for collagen fibrillogenesis and to interact with GAGs; thus, fragmented SLRP peptides generated from the cleavage of the core protein molecules interact with TLR-2, 4, TGFs, BMPs, MMPs, cytokines, and integrins during various stages of matrix dysregulation [7]. SLRP binding sites also provide adhesion for pathogens or toxic substances, such as in *Borrelia burgdorferi*, the causative agent of Lyme disease that resides in the extracellular matrix and binds with decorin via decorin-binding proteins A and B [88], and mice deficient in the decorin gene are more resistant to Lyme disease [89]. Multivalent binding abilities of SLRPs biglycan and decorin form complexes with low-density lipoprotein and apolipoproteins to collagen, which results in the accumulation of toxic metabolites in atherosclerotic plaque [90, 91]. Biglycan triggers cell-mediated inflammation by influencing the acitivity of chemokines, while B-cell activation is characterized as a multi-receptor signaling complex [15], leading to many other effects such as caspase activation, generation of catabolic cytokines, TNF-alpha, and IL-1beta [81]. SLRP fragments (Fig. 8.1) may be more potent in cell signalling than the intact molecules due to their smaller sizes and ability to internalize within the cells, thus leading to either proliferative or antiproliferative effects in various tissue types and cells [92] as discussed in the previous chapters. SLRPs thus exhibit diverse functions acting as modulators of tissue organization, cellular proliferation, matrix adhesion, and in integrating the cellular responses to growth factors and cytokines. SLRPs have emerging roles in health and disease, and are recognized as a key signaling molecules, along with growth factors, cytokines, and cell receptors that regulate cell growth, morphogenesis, and immunity.

8.5 Conclusions

SLRPs are reported to become fragmented into smaller sizes in cancerous tissue, articular cartilages, intervertebral discs, meniscus, and facet joints through the combined action of increased matrix catabolism/dysregulation, mechanical loading, and matrix-degrading enzymes. SLRP fragments thus represent a complex yet unidentified stage of various tissues in diseases. SLRPs are pericellular or territorial in localization and potentially act as DAMPs by interacting with cell surface pattern recognition receptors such as Toll-like receptors. Many past investigations, including reports from our lab, show that SLRP fragments are capable of interacting with cell surface receptors, such as integrins $\alpha_5\beta_1$, and initiate the NF-κB pain signaling pathways. Therefore, understanding the basis behind SLRPs fragments will likely lead to interventions that allow early diagnosis of degenerative changes as biomarkers of diseases, and analysis of the involved target pathways can lead to the development of therapeutics that can prevent and retard the process of pathological changes in patients.

References

1. Yang D, Han Z, Oppenheim JJ. Alarmins and immunity. Immunol Rev. 2017;280(1):41–56.
2. Gallucci S, Matzinger P. Danger signals: SOS to the immune system. Curr Opin Immunol. 2001;13(1):114–9.
3. Nathan C. Points of control in inflammation. Nature. 2002;420(6917):846–52.
4. Bianchi ME. DAMPs, PAMPs and alarmins: all we need to know about danger. J Leukoc Biol. 2007;81(1):1–5.
5. Chen GY, Nuñez G. Sterile inflammation: sensing and reacting to damage. Nat Rev Immunol. 2010;10(12):826–37.
6. Chen S, Birk DE. The regulatory roles of small leucine-rich proteoglycans in extracellular matrix assembly. FEBS J. 2013;280(10):2120–37.

7. Matsushima N, Miyashita H, Kretsinger RH. Sequence features, structure, ligand interaction, and diseases in small leucine rich repeat proteoglycans. J Cell Commun Signal. 2021;15(4):519–31.
8. Pellegata NS, Dieguez-Lucena JL, Joensuu T, Lau S, Montgomery KT, Krahe R, et al. Mutations in KERA, encoding keratocan, cause cornea plana. Nat Genet. 2000;25(1):91–5.
9. Silva T, Gomes CP, Voigt DD, de Souza RB, Medeiros K, Cosentino NL, et al. Fibromodulin gene variants (FMOD) as potential biomarkers for prostate cancer and benign prostatic hyperplasia. Disease Marker. 2022;2022:5215247.
10. Grover J, Chen XN, Korenberg JR, Recklies AD, Roughley PJ. The gene organization, chromosome location, and expression of a 55-kDa matrix protein (PRELP) of human articular cartilage. Genomics. 1996;38(2):109–17.
11. Le Goff MM, Hindson VJ, Jowitt TA, Scott PG, Bishop PN. Characterization of opticin and evidence of stable dimerization in solution. J Biol Chem. 2003;278(46):45280–7.
12. Kao WW, Liu CY. Roles of lumican and keratocan on corneal transparency. Glycoconj J. 2002;19(4–5):275–85.
13. Geng Y, McQuillan D, Roughley PJ. SLRP interaction can protect collagen fibrils from cleavage by collagenases. Matrix Biol. 2006;25(8):484–91.
14. Oldberg A, Antonsson P, Moses J, Fransson LA. Amino-terminal deletions in the decorin core protein leads to the biosynthesis of proteoglycans with shorter glycosaminoglycan chains. FEBS Lett. 1996;386(1):29–32.
15. Nastase MV, Young MF, Schaefer L. Biglycan: a multivalent proteoglycan providing structure and signals. J Histochem Cytochem. 2012;60(12):963–75.
16. Ameye L, Young MF. Mice deficient in small leucine-rich proteoglycans: novel in vivo models for osteoporosis, osteoarthritis, Ehlers-Danlos syndrome, muscular dystrophy, and corneal diseases. Glycobiology. 2002;12(9):107R–16R.
17. Chakravarti S, Magnuson T, Lass JH, Jepsen KJ, LaMantia C, Carroll H. Lumican regulates collagen fibril assembly: skin fragility and corneal opacity in the absence of lumican. J Cell Biol. 1998;141(5):1277–86.
18. Matheson S, Larjava H, Häkkinen L. Distinctive localization and function for lumican, fibromodulin and decorin to regulate collagen fibril organization in periodontal tissues. J Periodontal Res. 2005;40(4):312–24.
19. Baghy K, Dezso K, László V, Fullár A, Péterfia B, Paku S, et al. Ablation of the decorin gene enhances experimental hepatic fibrosis and impairs hepatic healing in mice. Lab Investig. 2011;91(3):439–51.
20. Wu Z, Aron AW, Macksoud EE, Iozzo RV, Hai CM, Lechner BE. Uterine dysfunction in biglycan and decorin deficient mice leads to dystocia during parturition. PLoS One. 2012;7(1):e29627.
21. Xu T, Bianco P, Fisher LW, Longenecker G, Smith E, Goldstein S, et al. Targeted disruption of the biglycan gene leads to an osteoporosis-like phenotype in mice. Nat Genet. 1998;20(1):78–82.
22. Ameye L, Aria D, Jepsen K, Oldberg A, Xu T, Young MF. Abnormal collagen fibrils in tendons of biglycan/fibromodulin-deficient mice lead to gait impairment, ectopic ossification, and osteoarthritis. FASEB J. 2002;16(7):673–80.
23. Corsi A, Xu T, Chen XD, Boyde A, Liang J, Mankani M, et al. Phenotypic effects of biglycan deficiency are linked to collagen fibril abnormalities, are synergized by decorin deficiency, and mimic Ehlers-Danlos-like changes in bone and other connective tissues. J Bone Miner Res. 2002;17(7):1180–9.
24. Jepsen KJ, Wu F, Peragallo JH, Paul J, Roberts L, Ezura Y, et al. A syndrome of joint laxity and impaired tendon integrity in lumican- and fibromodulin-deficient mice. J Biol Chem. 2002;277(38):35532–40.
25. Raghunath M, Superti-Furga A, Godfrey M, Steinmann B. Decreased extracellular deposition of fibrillin and decorin in neonatal Marfan syndrome fibroblasts. Hum Genet. 1993;90(5):511–5.

26. Izumi T, Tajima S, Nishikawa T. Stimulated expression of decorin and the decorin gene in fibroblasts cultured from patients with localized scleroderma. Arch Dermatol Res. 1995;287(5):417–20.
27. Beavan LA, Quentin-Hoffmann E, Schönherr E, Snigula F, Leroy JG, Kresse H. Deficient expression of decorin in infantile progeroid patients. J Biol Chem. 1993;268(13):9856–62.
28. Fedarko NS, Robey PG, Vetter UK. Extracellular matrix stoichiometry in osteoblasts from patients with osteogenesis imperfecta. J Bone Miner Res. 1995;10(7):1122–9.
29. Gu J, Wada Y. Aberrant expressions of decorin and biglycan genes in the carbohydrate-deficient glycoprotein syndrome. J Biochem. 1995;117(6):1276–9.
30. Raspanti M, Viola M, Forlino A, Tenni R, Gruppi C, Tira ME. Glycosaminoglycans show a specific periodic interaction with type I collagen fibrils. J Struct Biol. 2008;164(1):134–9.
31. Merline R, Schaefer RM, Schaefer L. The matricellular functions of small leucine-rich proteoglycans (SLRPs). J Cell Commun Signal. 2009;3(3–4):323–35.
32. Moffatt P, Geng Y, Lamplugh L, Nanci A, Roughley PJ. Absence of the dermatan sulfate chain of decorin does not affect mouse development. J Negat Results Biomed. 2017;16(1):7.
33. Seidler DG, Faiyaz-Ul-Haque M, Hansen U, Yip GW, Zaidi SH, Teebi AS, et al. Defective glycosylation of decorin and biglycan, altered collagen structure, and abnormal phenotype of the skin fibroblasts of an Ehlers-Danlos syndrome patient carrying the novel Arg270Cys substitution in galactosyltransferase I (beta4GalT-7). J Mol Med (Berl). 2006;84(7):583–94.
34. Choi RK, Smith MM, Martin JH, Clarke JL, Dart AJ, Little CB, et al. Chondroitin sulphate glycosaminoglycans contribute to widespread inferior biomechanics in tendon after focal injury. J Biomech. 2016;49(13):2694–701.
35. Rigozzi S, Müller R, Stemmer A, Snedeker JG. Tendon glycosaminoglycan proteoglycan sidechains promote collagen fibril sliding-AFM observations at the nanoscale. J Biomech. 2013;46(4):813–8.
36. Tatara Y, Suto S, Itoh K. Novel roles of glycosaminoglycans in the degradation of type I collagen by cathepsin K. Glycobiology. 2017;27(12):1089–98.
37. Johnstone B, Markopoulos M, Neame P, Caterson B. Identification and characterization of glycanated and non-glycanated forms of biglycan and decorin in the human intervertebral disc. Biochem J. 1993;292:661–6.
38. Waddington RJ, Hall RC, Embery G, Lloyd DM. Changing profiles of proteoglycans in the transition of predentine to dentine. Matrix Biol. 2003;22(2):153–61.
39. Sugars RV, Olsson ML, Marchner S, Hultenby K, Wendel M. The glycosylation profile of osteoadherin alters during endochondral bone formation. Bone. 2013;53(2):459–67.
40. Melching LI, Roughley PJ. Modulation of keratan sulfate synthesis on lumican by the action of cytokines on human articular chondrocytes. Matrix Biol. 1999;18(4):381–90.
41. Ernst S, Langer R, Cooney CL, Sasisekharan R. Enzymatic degradation of glycosaminoglycans. Crit Rev Biochem Mol Biol. 1995;30(5):387–444.
42. Iozzo RV. Matrix proteoglycans: from molecular design to cellular function. Annu Rev Biochem. 1998;67:609–52.
43. Imai K, Hiramatsu A, Fukushima D, Pierschbacher MD, Okada Y. Degradation of decorin by matrix metalloproteinases: identification of the cleavage sites, kinetic analyses and transforming growth factor- beta1 release. Biochem J. 1997;322(Pt 3):809–14.
44. Li Y, Aoki T, Mori Y, Ahmad M, Miyamori H, Takino T, et al. Cleavage of lumican by membrane- type matrix metalloproteinase-1 abrogates this proteoglycan-mediated suppression of tumor cell colony formation in soft agar. Cancer Res. 2004;64(19):7058–64.
45. Heathfield TF, Onnerfjord P, Dahlberg L, Heinegård D. Cleavage of fibromodulin in cartilage explants involves removal of the N-terminal tyrosine sulfate-rich region by proteolysis at a site that is sensitive to matrix metalloproteinase-13. J Biol Chem. 2004;279(8):6286–95.
46. Monfort J, Tardif G, Reboul P, Mineau F, Roughley P, Pelletier JP, et al. Degradation of small leucine-rich repeat proteoglycans by matrix metalloprotease-13: identification of a new biglycan cleavage site. Arthritis Res Ther. 2006;8(1):R26.

47. Zhang L, Yang M, Yang D, Cavey G, Davidson P, Gibson G. Molecular interactions of MMP-13 C-terminal domain with chondrocyte proteins. Connect Tissue Res. 2010;51(3):230–9.
48. Lama P, Claireaux H, Flower L, Harding IJ, Dolan T, Le Maitre CL, et al. Physical disruption of intervertebral disc promotes cell clustering and a degenerative phenotype. Cell Death Discov. 2019;5:154.
49. Bisson G, et al. Facet joint degeneration in adolescent idiopathic scoliosis. JOR Spine. 2018;1(2):e1016.
50. Lama P, Le Maitre CL, Dolan P, Tarlton JF, Harding IJ, Adams MA. Do intervertebral discs degenerate before they herniate, or after? Bone Joint J. 2013;95(8):1127–33.
51. Rees SG, Flannery CR, Little CB, Hughes CE, Caterson B, Dent CM. Catabolism of aggrecan, decorin and biglycan in tendon. Biochem J. 2000;350(Pt 1):181–8.
52. Egeblad M, Werb Z. New functions for the matrix metalloproteinases in cancer progression. Nat Rev Cancer. 2002;2(3):161–74.
53. Bian Y, Xiang Z, Wang Y, Ren Q, Chen G, Xiang B, et al. Immunomodulatory roles of metalloproteinases in rheumatoid arthritis. Front Pharmacol. 2023;14:1285455.
54. Boivin WA, Shackleford M, Vanden Hoek A, Zhao H, Hackett TL, Knight DA, et al. Granzyme B cleaves decorin, biglycan and soluble betaglycan, releasing active transforming growth factor-β1. PLoS One. 2012;7(3):e33163.
55. Tío L, Martel-Pelletier J, Pelletier JP, Bishop PN, Roughley P, Farran A, et al. Characterization of opticin digestion by proteases involved in osteoarthritis development. Joint Bone Spine. 2014;81(2):137–41.
56. Melching LI, Fisher WD, Lee ER, Mort JS, Roughley PJ. The cleavage of biglycan by aggrecanases. Osteoarthr Cartil. 2006;14(11):1147–54.
57. Haglund L, Tillgren V, Addis L, Wenglén C, Recklies A, Heinegård D. Identification and characterization of the integrin alpha2beta1 binding motif in chondroadherin mediating cell attachment. J Biol Chem. 2011;286(5):3925–34.
58. Wiberg C, Hedbom E, Khairullina A, Lamandé SR, Oldberg A, Timpl R, et al. Biglycan and decorin bind close to the n-terminal region of the collagen VI triple helix. J Biol Chem. 2001;276(22):18947–52.
59. Barreto G, Soininen A, Ylinen P, Sandelin J, Konttinen YT, Nordström DC, et al. Soluble biglycan: a potential mediator of cartilage degradation in osteoarthritis. Arthritis Res Ther. 2015;17:379.
60. Li Q, Han B, Wang C, Tong W, Wei Y, Tseng WJ, et al. Mediation of Cartilage Matrix Degeneration and Fibrillation by Decorin in Post-traumatic Osteoarthritis. Arthritis Rheumatol. 2020;72(8):1266–77.
61. Dings MPG, Manoukian P, Waasdorp C, Quik JSE, Strijker M, Lodestijn SC, et al. Serum levels of iCAF-derived osteoglycin predict favorable outcome in pancreatic cancer. Int J Cancer. 2023;152(3):511–23.
62. Hsueh MF, Önnerfjord P, Kraus VB. Biomarkers and proteomic analysis of osteoarthritis. Matrix Biol. 2014;39:56–66.
63. Anderson NM, Simon MC. The tumor microenvironment. Curr Biol. 2020;30(16):R921–R5.
64. Flannery CR. Usurped SLRPs: novel arthritis biomarkers exposed by catabolism of small leucine- rich proteoglycans? Arthritis Res Ther. 2006;8(2):106.
65. Schaefer L. Small leucine-rich proteoglycans in kidney disease. J Am Soc Nephrol. 2011;22(7):1200–7.
66. Adams MA, Lama P, Zehra U, Dolan P. Why do some intervertebral discs degenerate, when others (in the same spine) do not? Clin Anat. 2015;28(2):195–204.
67. Lama P, Le Maitre CL, Harding IJ, Dolan P, Adams MA. Nerves and blood vessels in degenerated intervertebral discs are confined to physically disrupted tissue. J Anat. 2018;233(1):86–97.
68. Jian J, Zheng Z, Zhang K, Rackohn TM, Hsu C, Levin A, et al. Fibromodulin promoted in vitro and in vivo angiogenesis. Biochem Biophys Res Commun. 2013;436(3):530–5.
69. Vogl T, Eisenblätter M, Völler T, Zenker S, Hermann S, van Lent P, et al. Alarmin S100A8/S100A9 as a biomarker for molecular imaging of local inflammatory activity. Nat Commun. 2014;5:4593.

70. Kessenbrock K, Plaks V, Werb Z. Matrix metalloproteinases: regulators of the tumor microenvironment. Cell. 2010;141(1):52–67.
71. Karamanos NK, Theocharis AD, Piperigkou Z, Manou D, Passi A, Skandalis SS, et al. A guide to the composition and functions of the extracellular matrix. FEBS J. 2021;288(24):6850–912.
72. Moreth K, Iozzo RV, Schaefer L. Small leucine-rich proteoglycans orchestrate receptor crosstalk during inflammation. Cell Cycle. 2012;11(11):2084–91.
73. Garantziotis S, Savani RC. Proteoglycans in Toll-like receptor responses and innate immunity. Am J Physiol Cell Physiol. 2022;323(1):C202–C14.
74. Weber KT, Satoh S, Alipui DO, Virojanapa J, Levine M, Sison C, et al. Exploratory study for identifying systemic biomarkers that correlate with pain response in patients with intervertebral disc disorders. Immunol Res. 2015;63(1–3):170–80.
75. Hu L, Zang MD, Wang HX, Li JF, Su LP, Yan M, et al. Biglycan stimulates VEGF expression in endothelial cells by activating the TLR signaling pathway. Mol Oncol. 2016;10(9):1473–84.
76. Matsushima N, Tanaka T, Enkhbayar P, Mikami T, Taga M, Yamada K, et al. Comparative sequence analysis of leucine-rich repeats (LRRs) within vertebrate toll-like receptors. BMC Genomics. 2007;8:124.
77. Frey H, Schroeder N, Manon-Jensen T, Iozzo RV, Schaefer L. Biological interplay between proteoglycans and their innate immune receptors in inflammation. FEBS J. 2013;280(10):2165–79.
78. di Celle PF, Carbone A, Marchis D, Zhou D, Sozzani S, Zupo S, et al. Cytokine gene expression in B-cell chronic lymphocytic leukemia: evidence of constitutive interleukin-8 (IL-8) mRNA expression and secretion of biologically active IL-8 protein. Blood. 1994;84(1):220–8.
79. Akasaki Y, Alvarez-Garcia O, Saito M, Caramés B, Iwamoto Y, Lotz MK. FoxO transcription factors support oxidative stress resistance in human chondrocytes. Arthritis Rheumatol. 2014;66(12):3349–58.
80. Haglund L, Bernier SM, Onnerfjord P, Recklies AD. Proteomic analysis of the LPS-induced stress response in rat chondrocytes reveals induction of innate immune response components in articular cartilage. Matrix Biol. 2008;27(2):107–18.
81. Schaefer L, Babelova A, Kiss E, Hausser HJ, Baliova M, Krzyzankova M, et al. The matrix component biglycan is proinflammatory and signals through Toll-like receptors 4 and 2 in macrophages. J Clin Invest. 2005;115(8):2223–33.
82. Xie C, Mondal DK, Ulas M, Neill T, Iozzo RV. Oncosuppressive roles of decorin through regulation of multiple receptors and diverse signaling pathways. Am J Physiol Cell Physiol. 2022;322(3):C554–C66.
83. Javaid N, Choi S. Toll-like Receptors from the Perspective of Cancer Treatment. Cancers (Basel). 2020;12(2)
84. Krock E, Rosenzweig DH, Currie JB, Bisson DG, Ouellet JA, Haglund L. Toll-like receptor activation induces degeneration of human intervertebral discs. Sci Rep. 2017;7(1):17184.
85. Lacagnina MJ, Watkins LR, Grace PM. Toll-like receptors and their role in persistent pain. Pharmacol Ther. 2018;184:145–58.
86. Phillips KL, Chiverton N, Michael AL, Cole AA, Breakwell LM, Haddock G, et al. The cytokine and chemokine expression profile of nucleus pulposus cells: implications for degeneration and regeneration of the intervertebral disc. Arthritis Res Ther. 2013;15(6):R213.
87. Ramesh G, MacLean AG, Philipp MT. Cytokines and chemokines at the crossroads of neuroinflammation, neurodegeneration, and neuropathic pain. Mediat Inflamm. 2013;2013:480739.
88. Gubbiotti MA, Vallet SD, Ricard-Blum S, Iozzo RV. Decorin interacting network: A comprehensive analysis of decorin-binding partners and their versatile functions. Matrix Biol. 2016;55:7–21.
89. Brown EL, Wooten RM, Johnson BJ, Iozzo RV, Smith A, Dolan MC, et al. Resistance to Lyme disease in decorin-deficient mice. J Clin Invest. 2001;107(7):845–52.
90. Kovanen PT, Pentikäinen MO. Decorin links low-density lipoproteins (LDL) to collagen: a novel mechanism for retention of LDL in the atherosclerotic plaque. Trends Cardiovasc Med. 1999;9(3–4):86–91.

91. Zeng X, Chen J, Miller YI, Javaherian K, Moulton KS. Endostatin binds biglycan and LDL and interferes with LDL retention to the subendothelial matrix during atherosclerosis. J Lipid Res. 2005;46(9):1849–59.
92. Iozzo RV, Schaefer L. Proteoglycans in health and disease: novel regulatory signaling mechanisms evoked by the small leucine-rich proteoglycans. FEBS J. 2010;277(19):3864–75.

SLRPs and Their Therapeutic Relevance

9

Polly Lama

Abstract

The SLRP core protein molecule and its fragmented peptides act to inhibit or activate various disease processes, regulating or altering the synthesis of several intercellular mechanisms, notably by increasing the expression of matrix-degrading enzymes and inflammatory mediators. These changes can lead to either enhanced cell proliferation or conversely trigger apoptosis. Among all the SLRPs, the most potent molecules exhibiting therapeutic relevance are decorin, lumican, fibromodulin core protein, along with their peptide derivatives. Recent and past studies show that almost all classes of SLRPs interact directly with the cell surface Toll-like receptors 2 and 4, mainly to downregulate or upregulate several signalling pathways that can impact the catalytic domain binding sites of matrix-degrading enzymes during diseases and disorders. Biglycan, in particular, accelerates inflammatory pathways involving TLR-2 and 4, and the purinergic P2X7 receptors; thus, biglycan is suggested to act as a potent damage-associated molecular pattern recognition molecule and serves as a biomarker of various diseases in the matrix, including cancer. Therefore, based on the diverse functions of SLRPs as described in the preceding chapters, the main objective of the current section is to summarize the potential role of SLRPs in therapeutic interventions.

Keywords

Small Leucine Rich Proteoglycans · Diseases · Therapeutics · Glycosaminoglycans · Decorin · Biglycan

P. Lama (✉)
Department of Anatomy, Sikkim Manipal Institute of Medical Sciences, Sikkim Manipal University, Gangtok, Sikkim, India
e-mail: polly.lama@smims.smu.edu.in

P. Lama (ed.), *The Biological Role of Small Leucine-Rich Proteoglycans (SLRPs) in Matrix Homeostasis and Diseases*,
https://doi.org/10.1007/978-981-96-7890-7_9

9.1 Decorin and Its Relevance in Cancer Therapeutics

Decorin and its proteolytic fragments *(as discussed in previous chapters)* act to inhibit tumorigenesis by promoting the release of anti-angiogenic factors, and decorin is also regarded as "a guardian from the matrix" [1]. Specific peptides of the decorin molecule act as a biomarkers in ischemic stroke [2], renal [3], and lung diseases [4]. Further, in patients with invasive and metastatic breast cancer, decorin has an oncosuppressive function, and loss of decorin is considered as an important clinical prognostic feature [5], which associates with the breast tumor stages [6]. Similarly, the expression of decorin decreases in urinary bladder carcinoma of varying grades [7], including plasma of patients with multiple myeloma [8], esophageal squamous cell carcinoma [9], and colon cancer [10]. Further, in solid malignancies involving the lungs, breast, and colon, the stromal tissue shows low levels of decorin with an in silico-based bioinformatics analysis [11]. Thus, stromally redeposited decorin in colon and breast carcinomas has the potential to suppress cell growth by restraining the tumor microenvironment in a paracrine mode that eventually decreases the expression of receptor tyrosine kinases present on the tumor cells [12]. Moreover, progressive depletion of decorin levels in prostate cancer, hepatic carcinomas, and differentiated sarcomas correlates with poor prognosis of the disease [13]. Therefore, tumor growth is directly linked to decorin levels thus positioning it as a potent biomarker in various cancer types.

9.1.1 Leucine-Rich Repeats (LRR) of Decorin

Leucine-rich repeats (LRR) of decorin, unlike the other SLRPs, present an interface capable of coordinating protein-protein interactions mediated by amino acid residues attached to the core protein structure [14]. Each of the LRRs attached to the decorin core protein molecule imparts functional properties that regulate the binding of decorin to connective tissue growth factors [15], transforming growth factor beta receptors, vascular endothelial growth factor receptor-2 [16], and the collagen binding sequence (SYIRIADTNIT) present on the solenoid curve of the LRR structure [17], allowing binding between decorin and type I collagen, and alterations in the decorin LRR structure tend to compromise the stability and arrangment of the collagen fiber in the matrix. Another significance of decorin LRRs is the presence of an elongated "ear repeat," a feature exhibited by SLRP members belonging to classes I, II, and III (*as described in* Chap. 1). Truncation of the ear repeats in the decorin protein causes congenital stromal corneal dystrophy due to frameshift mutation in the decorin gene structure [18]. Further, in an experimental mouse model with a lack of ear repeats revealed intracellular migration, accumulation, and deposition of decorin within the endoplasmic reticulum, and this may be attributed to extremely small molecular mass of decorin being less than 30 kDa. The internalization of decorin molecule causes corneal fibroblast cells to dysregulate collagen synthesis and assembly of fibrils ultimately leading to the reorganization of the matrix [19]. Decorin and its LRRs are thus involved in direct regulation of collagen synthesis and are proposed to act as a therapeutic target in corneal diseases.

9.1.2 Glycosaminoglycans (GAGs) Side Chain of Decorin

Glycosaminoglycans (GAGs) side chain of decorin exhibits functional changes in various types of pathologies associated with connective tissue [20], and gene mutations in the decorin GAG side chain lead to congenital stromal corneal dystrophy [21] Ehlers-Danlos syndrome [22], including lung and bladder cancer [23]. Therefore, improperly folded or missing GAG chain sequence disrupts the conformational rigidity between decorin and collagen fibrils, and in Ehlers-Danlos syndrome, this change is expressed as a skin fragility phenotype where secreted decorin lacks the GAG side chains [24]. Further, disruptions in the alignment of decorin and glycosaminoglycan side chain impairs collagen fibril formations that directly impacts the spatial composition and position of the collagen fibers in colon, ovarian, pancreatic and gastric cancer, and often in these cases, the chondroitin sulfate GAG side chain can predominant, while the dermatan sulfate side chain of decorin declines; additionally, it also has been demonstrated that the presence of chondroitin sulfate side chain allows cell migration, thereby contributing to the malignancy of the tumor [25]. Thus, the side chains of decorin play a regulatory role in disease progression making it a valuable biomarker for various pathological conditions.

9.1.3 Decorin and Its Fragmented Peptides

Decorin and its fragmented peptides interact with the Toll-like and met receptors in innate and adaptive inflammatory responses [26] and can act as a damage-associated molecular pattern recognition molecule [27] (Fig. 9.1). Pro-inflammatory interactions is modulated by circulating decorin and biglycan engaging TLR-2/4 receptors, increasing the synthesis and secretion of catabolic factors, such as TNF-α and IL-12 [28]. Decorin is therefore suggested to act as a pro-inflammatory molecule, a biomarker (*as already discussed in* Sect. 9.1), in various types of cancers, and decrease in the circulating level of decorin is suggested to be a potent target for therapeutic intervention [29].

Decorin and their peptides derivatives can also accelerate degradation of certain oncoproteins by inhibiting angiogenesis through the Met signalling pathway and the transient receptor tyrosine complexes. Analysis with phosphotyrosine RTK array shows Met receptors are activated by the soluble fragment of decorin as well as the protein core molecule [26]. The tight binding between Met and decorin facilitates anti-tumorigenesis activity limiting the growth of blood vessels and even promotes pro-mitophagic functions by competing with the epidermal growth factor receptor, which is the primary initiator of mitophagy [28]. Consequently, Met binding to decorin antagonizes two oncogenes, β-catenin and Myc, via the 26S proteasome

SLRP's	Chondroadherin (CHAD)	Decorin (DCN)	Biglycan (BGN)	Fibromodulin (FMOD)	Lumican (LUM)
Size of core protein in IVDs, knee meniscus & articular cartilage	38, 36 KDa	43, 45, 50 KDa	42, 45 and 48 KDa reported	42, 45 and (50 KDa reported in IVDs)	40,45,48 KDa
Fragments in IVDs, knee meniscus, articular cartilage	35, 28	37, 29, 18	39, 38, 36, 28, 27	36, 38	36, 28, 27
Fragments in Adolescent idiopathic Scoliosis articular facets	35 to 22 kDa multiple bands	36 to 22 kDa	36 to 22 kDa	36 kDa, few fragments	Non-fragmented

Fig. 9.1 Proteolytic SLRP fragments and core protein of chondroadherin, decorin, biglycan, fibromodulin, and lumican in intervertebral disc, knee meniscus, articular cartilage, and adolescent idiopathic scoliosis articular facets. (Figure adapted from Dr. Polly Lama, Molecular Matrix Lab)

[26, 30] suppressing β-catenin signalling. In tumors, direct phosphorylation of β-catenin by RTKs is suggested to promote cell invasions in various cancer tissues [29, 31]. Similarly, Met activation in colon carcinomas also directly influences β-catenin signalling, where loss of decorin potentially alters the action on RTKs, contributing toward transformed Met/β catenin signalling in the lining epithelium of intestine contributing toward malignancies. Further, decorin and Met binding suppress angiogenesis in cervical and breast carcinomas [5] by enhancing the expression and secretion of antiangiogenic factors, such as tissue inhibitor of matrix metalloproteinases (TIMP-3) and thrombospondin motif (TSP-1) [32, 33]. Recent investigation also indicates that decorin facilitates secretion of thrombospondin-1 in MDA-MB-231 breast cancer cell line, inhibiting endothelial blood vessels infiltration, and this activity is comparable to the response as mediated by epidermal growth factor receptors that are known to regulate antiangiogenic activity by modulating the RhoA/ROCK-1 signalling pathways in the control cells [33, 34]. Therefore, the antiangiogenic activity of TIMP-3 and TSP-1 and their involvement in several pathophysiological processes are associated with the availability of decorin in the malignant cells and possibly play a role in arresting cancer cell growth and metabolism. Therefore, decorin and its peptides regulate angiogenesis by silencing the mechanisms behind tumor neovascularization and decorin validates itself as an anti-tumorigenic molecule that is capable of regulating anti-metastatic functions.

9.1.4 Preclinical Role of Decorin Gene via the Adenovirus Vectors

Preclinical role of decorin gene via *the adenovirus vectors* through systemic administration of decorin core protein has been explored in the treatment of cancers [30, 34]. Although the therapeutic efficacy has remained varied in totally eliminating the disease, decorin protein core and its peptides arrest the growth of tumor cells by triggering apoptotic cell death by caspase-3 activation, hindering angiogenesis, and further preventing metastases in various types of orthotopic tumor xenografts [35, 36]. Collectively, many such previously published findings demonstrates the in vivo efficacy of decorin as a therapeutic agent by suppressing the oncogenic and angiogenic traits that occur during the formation of tumors.

9.2 Biglycan

Biglycan show structural and functional similarities with decorin, and both of these SLRPs comprise of leucine-rich repeats flanked by cysteine-rich domains (*discussed in detail in* Chaps. 1, 2, and 3). Biglycan is abundantly located in cartilages, tendon, and muscle tissues; thus, changes in the expression of biglycan level have been considered to be a potential therapeutic target in musculoskeletal disorders, and like decorin, one of the prime functions of biglycan is to regulate the process of collagen fibrillogenesis essential for maintaining the organization of muscles, bones, and cartilage tissues [28, 37].

9.2.1 Biglycan in Bone Repair and Regeneration

Biglycan in bone repair and regeneration acts by promoting the function of bone morphogenetic protein BMP-2/4, favoring osteogenesis (a process in bone formation), and mice deficient in the biglycan gene exhibit shorter bones (osteopenia) with decreased bone mass (osteoporosis) in comparison to normal mice [38]. As discussed in the previous chapters, biglycan is capable of regulating multiple cell signalling pathways critical to bone metabolism, and in humans, this process is regulated by maintaining the homeostatic balance between osteoblasts, a type of bone forming cells, and osteoclasts, another type of bone removing cells, and biglycan modulates the controlled expression of anabolic mediators in these cells, especially during bone formation and resorption. Thus, (A) absence of biglycan during bone formation directly decreases osteoblastic cells [39], and recent analysis further shows biglycan compromises the synthesis of bone morphogenic protein 2/4 (BMP-2/4), which is a well-known inducer of osteogenesis; therefore, BMPs are currently being used therapeutically to aid bone repair. Osteoblast cells without the biglycan gene exhibit a loss in BMP-2/4 binding that subsequently culminates as a loss in osteogenic differentiation of bone cells. Therefore, biglycan and BMP-2/4 are essential factors that modulate bone regeneration, and repair process, especially during traumatic fractures [40].

9.2.2 Biglycan Promotes the Differentiation of Cellular Tendinous Matrix

Biglycan promotes the differentiation of cellular tendinous matrix by modulating the expression of signalling pathways mediated by Wnt, connective growth factors, and basic fibroblast growth factors [41]. In tendons, a population of tendon-specific stem/progenitor cell source differentiates to control the regenerative capabilities during injuries, and targeted deletion of biglycan and fibromodulin gene impedes the synthesis of bone morphogenetic protein-2 that impacts the differentiating abilities of tendon stem/progenitor cells to form ectopic bones instead of tenocytes impairing normal movements in mice [41, 42]. Similarly in human skeletal system biglycan is involved in the mutation of Wnt signalling pathway through the low density lipoprotein related receptor (LRP-6) causing several types of skeletal defects including spina bifida, elevated cholesterol level, and severe loss of bone mass or osteopenia, therefore exogenous application of non-glycanated biglycan restores the defective Wnt signalling in cell lines expressing mutant low density lipoprotein receptor-6 [43]. Thus, biglycan in bones, skeletal system, and tendons potentially treats pathologies caused by dysregulated Wnt signalling, and in tendons, biglycan additionally upregulates the synthesis of growth factors essential for differentiation of tendon-specific stem cells.

9.2.3 Biglycan Organizes Muscle Microstructure

Biglycan organizes muscle microstructure at the cell membranes required for initiation of mechanotransduction events to counteract the micro-shear forces applied to the cells and thereby protects the cell from abnormal biomechanical stress, essential for maintaining normal functions in muscles. Dystrophin and utrophin-membrane glycoproteins, also called DGC and UGC complexes, links muscle cell cytoskeleton to the surrounding extracellular matrix, capable of assisting several stress-related signaling events in the muscle tissues, and biglycan binds to the DGC and UGC involving the peripheral membrane protein α-dystroglycan, transmembrane protein α-sarcoglycan, and γ-sarcoglycan [44, 45]. Therefore, defects in these interactions result in Duchenne muscular dystrophy (DMD), and application of recombinant non-glycanated biglycan can restore the expression of several of these named components in the muscular cell membranes [46]. DMD is shown to result from mutations in dystrophin, a large intracellular protein that links the actin cytoskeleton to the membrane that anchors the DGC, and a non-glycanated form of biglycan [47], termed TVN-102 is currently under investigation for preclinical application in DMD [48].

9.2.4 Biglycan Monitors Synaptic Function in Neurodegenerative Disorders

Biglycan monitors synaptic function in neurodegenerative disorders by binding to the tyrosine kinase receptor family, termed the MuSK receptor, that regulates synaptic functions, including differentiation and maintenance of neural tissue [49]. Biglycan in an animal model regulates the expression of MuSK receptors in vivo and plays a role in maintaining synaptic stability, which is observed in almost all neurodegenerative diseases, including amyotrophic lateral sclerosis and spinal muscular atrophy. Thus, treatments promoting neuromuscular synapse by utilizing biglycan are a proposed therapeutic strategy that could potentially improve survival in patients with motor neuron diseases [49].

9.2.5 Biglycan as a Biomarker for Renal Diseases

Biglycan as a biomarker for renal diseases is proposed due to its extensive spatial localization within the human renal cortex, mainly confined to the tubulointerstitium tissue region, peritubular mesenchymal cells, distal collecting tubules, and endothelial cells of the kidney, while weaker spatial localization is observed in mesangial and epithelial cells of the glomeruli [50]. It has been demonstrated in rats that biglycan and decorin synthesis is induced by TGF-β, which was mainly detected in the cell-conditioned media rather than the tissues [51], indicating that biglycan and decorin are secreted by the kidney tissue. Biglycan, in addition, is also produced by glomerular epithelial cells when treated with TGF-β in cell culture [52], and therefore, in diabetic rats, mRNA levels of biglycan are elevated along with TGF-β [53]. Further, it has also been demonstrated, rats with renal transplant failure tend to develop autoantibodies against mesangial cells, forming focal adhesion plaques with vinculin, biglycan, and decorin [54]. Thus, these molecules contribute to the tissue damage and interfere with the repair process. In renal inflammation, biglycan forms an endogenous ligand with TLR-2 and 4 cell receptors, and soluble biglycan also acts as a damage-associated molecular pattern recognition molecule that triggers various pro-inflammatory signals in the adjacent organs, leading to further accumulation of biglycan [47, 55]. Overexpression of soluble biglycan has been reported in numerous diseases, and in kidney tissues, it aggravates the pathological progression of the disease [56] by playing a critical role in the regulation of inflammation [57, 58]. Therefore, preventing the release of soluble biglycan and blocking of biglycan-mediated Toll-like receptors 2 and 4 signalling could lead to the development of novel therapeutic targets in treating renal inflammation; alternately, the cleaved/fragmented form of biglycan released during the process of matrix remodeling also serves as a biomarker in the progression of renal diseases [47, 59]. Identification of biglycan peptides through a protein fingerprint approach is suggested to generate antibodies that specifically recognize biglycan peptides as neo-epitopes in biological fluids, including the plasma [60]. Thus, biglycan is suggested to act as an indicator of pathophysiological outcome in renal diseases (Fig. 9.1).

9.3 Lumican

Lumican is a suggested biomarker in various cancer pathologies, and it performs paradoxical function as an anticancer agent as well as a tumor promoter [61]. Lumican in cancer exhibits changes associated with tumorigenesis and is involved with the pathological progression of cancer cells that undergo epithelial-to-mesenchymal transition, thus leading to cell proliferation, migration, and adhesion, promoting immune cell invasions and uncontrolled inflammation that complicate the severity of the disease [62]. Lumican positively correlates with development of esophageal cancer [63], squamous cell carcinoma of lungs [64], lung adenocarcinoma [65], breast cancer [66], osteosarcoma [67], colorectal carcinoma [68], pancreatic [69], ovarian cancer [70], and gastric cancer [71] as described in *detail in* Chap. 2. In addition to its role in tumor growth, lumican also shows a contrasting characteristic, acting as an antitumor agent [72], suppressing the growth of pancreatic cancer cells [73]. In melanoma, lumican inhibits cancer cell migration by MMP-14 activity that controls cell proliferation by involving the Snail-1 transcription factor regulating epithelial to mesenchymal transition [74, 75], and a similar action of lumican is reported in the B16F1 cell line during lung metastasis nodule development [76] and prostate cancer [77]. Antitumor activity of lumican is suggested to stem from smaller fragments of lumican peptides that act as MMP inhibitors, as lumican peptides can interact with the integrin $\alpha 2\beta 1$ receptor, inhibiting neovascularization essential for tumor cell growth and nutrition [78]. Thus, application of recombinant lumican peptides on skin wounds of mice enhances healing as lumican promotes contractility of skin fibroblast cells by activating the integrin alpha-2 beta-1 receptors that facilitate closure of wounds. Similarly, when lumican is overexpressed through an adenoviral vector in a rabbit model, fibroblast cells thin out around the scarred tissue region by inhibiting cell proliferation, and which suppresses phosphorylation of focal adhesion kinase (FAK) binding with integrin alpha-2 beta-1 receptors involved in promoting cell adhesion [30]. Small peptides designed to promote the action of lumican favor corneal wound healing in mice involving transforming growth factor (TGF) beta-1 receptor complexes [79, 80]; thus, a lumican-derived peptide named lumcorin in melanoma shows therapeutic action by inhibiting cell chemotaxis and growth through the action of MMP-14 [78]. These findings therefore indicate biological activities of lumican in disease progression and inhibition, and lumican acts as a potential target in the treatment of various types of diseases and superficial injuries.

9.4 Fibromodulin

Fibromodulin protein and gene are potential biomarkers in benign or malignant prostatic cancers [81], and FMOD core protein and their small peptides are detected in circulating body fluids such as blood, urine, and prostatic secretions, including synovium, cartilages, and bones (Fig. 9.1). Fibromodulin gene expression is upregulated in malignant glioblastoma [82], and fibromodulin peptide acts as a novel tumor-associated antigen in lymphoma, leiomyoma, and leukemia, where FMOD allows expansions of CD8+ autologous T-lymphocytes in chronic lymphocytic leukemia [83, 84]. It has also been evidenced that the mRNA level of fibromodulin is upregulated in malignancies such as breast [85], prostate [86], and lung carcinomas [87]; thus, loss of fibromodulin in various experimental carcinomas decreases interstitial fluid pressure that results in loss of collagen fibril thickness, and as already described in the previous chapters, fibromodulin and other SLRP members act as a primary regulator of cell adhesion and collagen fibrillogenesis that contribute toward suppression of tumor growth by signalling transcription factors that prevent apoptosis [88, 89]. Thus, overexpression of fibromodulin in B-cell chronic lymphocytic leukemia is a target for therapeutic approaches [83], and inhibition of fibromodulin facilitates cell survival. Similarly, in glioblastoma, fibromodulin enables cell migration, leading to the activation of filamentous actin stress fiber, and an inhibitor-based 9-mer wild-type fibromodulin peptide, synthesized from the C-terminus of the protein core molecule, prevents the migration of these glioblastoma cells [82], further silencing FMOD gene associates with inhibition of TGF-β1-mediated glioblastoma cell migration. Thus, FMOD performs as a potential target in the treatment of various type of cancers.

Fibromodulin also plays a role in fetal-like scarless wound healing, promoting angiogenesis in skin tissue [90] by stimulating endothelial cell-derived capillary formations in chick chorioallantoic membrane assays, which even restores the vascularity of fibromodulin double knockout mouse wounds [91, 92]. These results suggest fibromodulin is a potent therapeutic candidate for wound management, especially in cases where angiogenesis is impaired, such as diabetic wounds.

Adenoviral transfection of a viral vector with the fibromodulin gene shows that it can also act as a target for wound healing in cancers [93], and in rats, recombinant adenoviral vector expressing fibromodulin in the treatment of streptozotocin-induced diabetic nephropathy shows downregulated expression of TGF beta1 [94]. Similarly, in rats, nonviral histidylated vectors with fibromodulin gene enhance tendon healing, including improvements of biomechanical and histological features in the achilles tendon injury model [95]. Thus, fibromodulin is also a potent target in diseases of the matrix.

9.5 SLRP Fragments in Cartilaginous Tissue

SLRP fragments in cartilaginous tissue are localized and extracted in various types of pathological conditions, *as described already in the previous chapters*, and biglycan core protein and their fragments are labeled as biomarkers of cartilage tissue degeneration and potent damage-associated molecular pattern recognition molecule; however, the therapeutic relevance of numerous SLRP fragments in cartilage tissue disorder is yet to be fully explored (Fig. 9.1).

References

1. Neill T, Schaefer L, Iozzo RV. Decorin: a guardian from the matrix. Am J Pathol. 2012;181(2):380–7.
2. Xu YZ, Yang ZG, Zhang YH, Zhang YW, Hong B, Liu JM. Dynamic reduction of plasma decorin following ischemic stroke: a pilot study. Neurochem Res. 2012;37(9):1843–8.
3. Schaefer L. Small leucine-rich proteoglycans in kidney disease. J Am Soc Nephrol. 2011;22(7):1200–7.
4. Shi X, Liang W, Yang W, Xia R, Song Y. Decorin is responsible for progression of non-small-cell lung cancer by promoting cell proliferation and metastasis. Tumour Biol. 2015;36(5):3345–54.
5. Goldoni S, Seidler DG, Heath J, Fassan M, Baffa R, Thakur ML, et al. An antimetastatic role for decorin in breast cancer. Am J Pathol. 2008;173(3):844–55.
6. Matsumine A, Shintani K, Kusuzaki K, Matsubara T, Satonaka H, Wakabayashi T, et al. Expression of decorin, a small leucine-rich proteoglycan, as a prognostic factor in soft tissue tumors. J Surg Oncol. 2007;96(5):411–8.
7. Sainio A, Nyman M, Lund R, Vuorikoski S, Boström P, Laato M, et al. Lack of decorin expression by human bladder cancer cells offers new tools in the therapy of urothelial malignancies. PLoS One. 2013;8(10):e76190.
8. Kristensen IB, Pedersen L, Rø TB, Christensen JH, Lyng MB, Rasmussen LM, et al. Decorin is down-regulated in multiple myeloma and MGUS bone marrow plasma and inhibits HGF-induced myeloma plasma cell viability and migration. Eur J Haematol. 2013;91(3):196–200.
9. Wu IC, Wu DC, Huang CC, Lin HS, Chen YK, Tsai HJ, et al. Plasma decorin predicts the presence of esophageal squamous cell carcinoma. Int J Cancer. 2010;127(9):2138–46.
10. Nyman MC, Sainio AO, Pennanen MM, Lund RJ, Vuorikoski S, Sundström JT, et al. Decorin in human colon cancer: localization in vivo and effect on cancer cell behavior in vitro. J Histochem Cytochem. 2015;63(9):710–20.
11. Bozoky B, Savchenko A, Guven H, Ponten F, Klein G, Szekely L. Decreased decorin expression in the tumor microenvironment. Cancer Med. 2014;3(3):485–91.
12. Neill T, Torres A, Buraschi S, Owens RT, Hoek JB, Baffa R, et al. Decorin induces mitophagy in breast carcinoma cells via peroxisome proliferator-activated receptor γ coactivator-1α (PGC-1α) and mitostatin. J Biol Chem. 2014;289(8):4952–68.
13. Järvinen TA, Prince S. Decorin: a growth factor antagonist for tumor growth inhibition. Biomed Res Int. 2015;2015:654765.
14. Enkhbayar P, Kamiya M, Osaki M, Matsumoto T, Matsushima N. Structural principles of leucine-rich repeat (LRR) proteins. Proteins. 2004;54(3):394–403.
15. Vial C, Gutiérrez J, Santander C, Cabrera D, Brandan E. Decorin interacts with connective tissue growth factor (CTGF)/CCN2 by LRR12 inhibiting its biological activity. J Biol Chem. 2011;286(27):24242–52.
16. Khan GA, Girish GV, Lala N, Di Guglielmo GM, Lala PK. Decorin is a novel VEGFR-2-binding antagonist for the human extravillous trophoblast. Mol Endocrinol. 2011;25(8):1431–43.

17. Kalamajski S, Aspberg A, Oldberg A. The decorin sequence SYIRIADTNIT binds collagen type I. J Biol Chem. 2007;282(22):16062–7.
18. Chen S, Sun M, Meng X, Iozzo RV, Kao WW, Birk DE. Pathophysiological mechanisms of autosomal dominant congenital stromal corneal dystrophy: C-terminal-truncated decorin results in abnormal matrix assembly and altered expression of small leucine-rich proteoglycans. Am J Pathol. 2011;179(5):2409–19.
19. Chen S, Sun M, Iozzo RV, Kao WW, Birk DE. Intracellularly-retained decorin lacking the C-terminal ear repeat causes ER stress: a cell-based etiological mechanism for congenital stromal corneal dystrophy. Am J Pathol. 2013;183(1):247–56.
20. Paganini C, Costantini R, Superti-Furga A, Rossi A. Bone and connective tissue disorders caused by defects in glycosaminoglycan biosynthesis: a panoramic view. FEBS J. 2019;286(15):3008–32.
21. Mizumoto S, Yamada S. Congenital disorders of deficiency in glycosaminoglycan biosynthesis. Front Genet. 2021;12:717535.
22. Seidler DG. The galactosaminoglycan-containing decorin and its impact on diseases. Curr Opin Struct Biol. 2012;22(5):578–82.
23. Wei J, Hu M, Huang K, Lin S, Du H. Roles of proteoglycans and Glycosaminoglycans in cancer development and progression. Int J Mol Sci. 2020;21(17)
24. Rühland C, Schönherr E, Robenek H, Hansen U, Iozzo RV, Bruckner P, et al. The glycosaminoglycan chain of decorin plays an important role in collagen fibril formation at the early stages of fibrillogenesis. FEBS J. 2007;274(16):4246–55.
25. Iozzo RV, Schaefer L. Proteoglycans in health and disease: novel regulatory signaling mechanisms evoked by the small leucine-rich proteoglycans. FEBS J. 2010;277(19):3864–75.
26. Goldoni S, Humphries A, Nyström A, Sattar S, Owens RT, McQuillan DJ, et al. Decorin is a novel antagonistic ligand of the met receptor. J Cell Biol. 2009;185(4):743–54.
27. Merline R, Moreth K, Beckmann J, Nastase MV, Zeng-Brouwers J, Tralhão JG, et al. Signaling by the matrix proteoglycan decorin controls inflammation and cancer through PDCD4 and MicroRNA-21. Sci Signal. 2011;4(199):ra75.
28. Nastase MV, Young MF, Schaefer L. Biglycan: a multivalent proteoglycan providing structure and signals. J Histochem Cytochem. 2012;60(12):963–75.
29. Buraschi S, Pal N, Tyler-Rubinstein N, Owens RT, Neill T, Iozzo RV. Decorin antagonizes met receptor activity and down-regulates {beta}-catenin and Myc levels. J Biol Chem. 2010;285(53):42075–85.
30. Tralhão JG, Schaefer L, Micegova M, Evaristo C, Schönherr E, Kayal S, et al. In vivo selective and distant killing of cancer cells using adenovirus-mediated decorin gene transfer. FASEB J. 2003;17(3):464–6.
31. Yang Y. Wnt signaling in development and disease. Cell Biosci. 2012;2(1):14.
32. Neill T, Painter H, Buraschi S, Owens RT, Lisanti MP, Schaefer L, et al. Decorin antagonizes the angiogenic network: concurrent inhibition of met, hypoxia inducible factor 1α, vascular endothelial growth factor a, and induction of thrombospondin-1 and TIMP3. J Biol Chem. 2012;287(8):5492–506.
33. Neill T, Jones HR, Crane-Smith Z, Owens RT, Schaefer L, Iozzo RV. Decorin induces rapid secretion of thrombospondin-1 in basal breast carcinoma cells via inhibition of Ras homolog gene family, member A/Rho-associated coiled-coil containing protein kinase 1. FEBS J. 2013;280(10):2353–68.
34. Neill T, Schaefer L, Iozzo RV. Decorin as a multivalent therapeutic agent against cancer. Adv Drug Deliv Rev. 2016;97:174–85.
35. Seidler DG, Goldoni S, Agnew C, Cardi C, Thakur ML, Owens RT, et al. Decorin protein core inhibits in vivo cancer growth and metabolism by hindering epidermal growth factor receptor function and triggering apoptosis via caspase-3 activation. J Biol Chem. 2006;281(36):26408–18.

36. Mondal DK, Xie C, Pascal GJ, Buraschi S, Iozzo RV. Decorin suppresses tumor lymphangiogenesis: a mechanism to curtail cancer progression. Proc Natl Acad Sci USA. 2024;121(18):e2317760121.
37. Young MF, Fallon JR. Biglycan: a promising new therapeutic for neuromuscular and musculoskeletal diseases. Curr Opin Genet Dev. 2012;22(4):398–400.
38. Xu T, Bianco P, Fisher LW, Longenecker G, Smith E, Goldstein S, et al. Targeted disruption of the biglycan gene leads to an osteoporosis-like phenotype in mice. Nat Genet. 1998;20(1):78–82.
39. Chen XD, Fisher LW, Robey PG, Young MF. The small leucine-rich proteoglycan biglycan modulates BMP-4-induced osteoblast differentiation. FASEB J. 2004;18(9):948–58.
40. Shainer R, Kram V, Kilts TM, Li L, Doyle AD, Shainer I, et al. Biglycan regulates bone development and regeneration. Front Physiol. 2023;14:1119368.
41. Bi Y, Ehirchiou D, Kilts TM, Inkson CA, Embree MC, Sonoyama W, et al. Identification of tendon stem/progenitor cells and the role of the extracellular matrix in their niche. Nat Med. 2007;13(10):1219–27.
42. Young MF, Bi Y, Ameye L, Chen XD. Biglycan knockout mice: new models for musculoskeletal diseases. Glycoconj J. 2002;19(4–5):257–62.
43. Berendsen AD, Fisher LW, Kilts TM, Owens RT, Robey PG, Gutkind JS, et al. Modulation of canonical Wnt signaling by the extracellular matrix component biglycan. Proc Natl Acad Sci USA. 2011;108(41):17022–7.
44. Bowe MA, Mendis DB, Fallon JR. The small leucine-rich repeat proteoglycan biglycan binds to alpha-dystroglycan and is upregulated in dystrophic muscle. J Cell Biol. 2000;148(4):801–10.
45. Rafii MS, Hagiwara H, Mercado ML, Seo NS, Xu T, Dugan T, et al. Biglycan binds to alpha- and gamma-sarcoglycan and regulates their expression during development. J Cell Physiol. 2006;209(2):439–47.
46. Mercado ML, Amenta AR, Hagiwara H, Rafii MS, Lechner BE, Owens RT, et al. Biglycan regulates the expression and sarcolemmal localization of dystrobrevin, syntrophin, and nNOS. FASEB J. 2006;20(10):1724–6.
47. Moreth K, Brodbeck R, Babelova A, Gretz N, Spieker T, Zeng-Brouwers J, et al. The proteoglycan biglycan regulates expression of the B cell chemoattractant CXCL13 and aggravates murine lupus nephritis. J Clin Invest. 2010;120(12):4251–72.
48. Fallon JR, McNally EM. Non-Glycanated Biglycan and LTBP4: leveraging the extracellular matrix for Duchenne muscular dystrophy therapeutics. Matrix Biol. 2018;68-69:616–27.
49. Amenta AR, Creely HE, Mercado ML, Hagiwara H, McKechnie BA, Lechner BE, et al. Biglycan is an extracellular MuSK binding protein important for synapse stability. J Neurosci. 2012;32(7):2324–34.
50. Schaefer L, Gröne HJ, Raslik I, Robenek H, Ugorcakova J, Budny S, et al. Small proteoglycans of normal adult human kidney: distinct expression patterns of decorin, biglycan, fibromodulin, and lumican. Kidney Int. 2000;58(4):1557–68.
51. Border WA, Noble NA, Yamamoto T, Harper JR, Yamaguchi Y, Pierschbacher MD, et al. Natural inhibitor of transforming growth factor-beta protects against scarring in experimental kidney disease. Nature. 1992;360(6402):361–4.
52. Nakamura T, Miller D, Ruoslahti E, Border WA. Production of extracellular matrix by glomerular epithelial cells is regulated by transforming growth factor-beta 1. Kidney Int. 1992;41(5):1213–21.
53. Yamamoto T, Noble NA, Miller DE, Border WA. Sustained expression of TGF-beta 1 underlies development of progressive kidney fibrosis. Kidney Int. 1994;45(3):916–27.
54. Paul LC, Muralidharan J, Muzaffar SA, Manting EH, Valentin JF, de Heer E, et al. Antibodies against mesangial cells and their secretory products in chronic renal allograft rejection in the rat. Am J Pathol. 1998;152(5):1209–23.
55. Zeng-Brouwers J, Beckmann J, Nastase MV, Iozzo RV, Schaefer L. De novo expression of circulating biglycan evokes an innate inflammatory tissue response via MyD88/TRIF pathways. Matrix Biol. 2014;35:132–42.

56. Babelova A, Moreth K, Tsalastra-Greul W, Zeng-Brouwers J, Eickelberg O, Young MF, et al. Biglycan, a danger signal that activates the NLRP3 inflammasome via toll-like and P2X receptors. J Biol Chem. 2009;284(36):24035–48.
57. Moreth K, Frey H, Hubo M, Zeng-Brouwers J, Nastase MV, Hsieh LT, et al. Biglycan-triggered TLR-2- and TLR-4-signaling exacerbates the pathophysiology of ischemic acute kidney injury. Matrix Biol. 2014;35:143–51.
58. Nastase MV, Iozzo RV, Schaefer L. Key roles for the small leucine-rich proteoglycans in renal and pulmonary pathophysiology. Biochim Biophys Acta. 2014;1840(8):2460–70.
59. Karsdal MA, Bay-Jensen AC, Leeming DJ, Henriksen K, Christiansen C. Quantification of "end products" of tissue destruction in inflammation may reflect convergence of cytokine and signaling pathways – implications for modern clinical chemistry. Biomarkers. 2013;18(5):375–8.
60. Genovese F, Manresa AA, Leeming DJ, Karsdal MA, Boor P. The extracellular matrix in the kidney: a source of novel non-invasive biomarkers of kidney fibrosis? Fibrogenesis Tissue Repair. 2014;7(1):4.
61. Giatagana EM, Berdiaki A, Tsatsakis A, Tzanakakis GN, Nikitovic D. Lumican in carcinogenesis-revisited. Biomol Ther. 2021;11:9.
62. Nikitovic D, Papoutsidakis A, Karamanos NK, Tzanakakis GN. Lumican affects tumor cell functions, tumor-ECM interactions, angiogenesis and inflammatory response. Matrix Biol. 2014;35:206–14.
63. Yamauchi N, Kanke Y, Saito K, Okayama H, Yamada S, Nakajima S, et al. Stromal expression of cancer-associated fibroblast-related molecules, versican and lumican, is strongly associated with worse relapse-free and overall survival times in patients with esophageal squamous cell carcinoma. Oncol Lett. 2021;21(6):445.
64. Matsuda Y, Yamamoto T, Kudo M, Kawahara K, Kawamoto M, Nakajima Y, et al. Expression and roles of lumican in lung adenocarcinoma and squamous cell carcinoma. Int J Oncol. 2008;33(6):1177–85.
65. Cappellesso R, Millioni R, Arrigoni G, Simonato F, Caroccia B, Iori E, et al. Lumican is overexpressed in lung adenocarcinoma pleural effusions. PLoS One. 2015;10(5):e0126458.
66. Karamanou K, Franchi M, Vynios D, Brézillon S. Epithelial-to-mesenchymal transition and invadopodia markers in breast cancer: Lumican a key regulator. Semin Cancer Biol. 2020;62:125–33.
67. Nikitovic D, Berdiaki A, Zafiropoulos A, Katonis P, Tsatsakis A, Karamanos NK, et al. Lumican expression is positively correlated with the differentiation and negatively with the growth of human osteosarcoma cells. FEBS J. 2008;275(2):350–61.
68. Alsheridah N, Akhtar S. Diet, obesity and colorectal carcinoma risk: results from a national cancer registry-based middle-eastern study. BMC Cancer. 2018;18(1):1227.
69. Ping LY, Ishiwata T, Asano G. Lumican expression in alpha cells of islets in pancreas and pancreatic cancer cells. J Pathol. 2002;196(3):324–30.
70. Zhang B, Cai FF, Zhong XY. An overview of biomarkers for the ovarian cancer diagnosis. Eur J Obstet Gynecol Reprod Biol. 2011;158(2):119–23.
71. Sun C, Chen Y, Kim NH, Lowe S, Ma S, Zhou Z, et al. Identification and verification of potential biomarkers in gastric cancer by integrated Bioinformatic analysis. Front Genet. 2022;13:911740.
72. Appunni S, Rubens M, Ramamoorthy V, Anand V, Khandelwal M, Saxena A, et al. Lumican, pro-tumorigenic or anti-tumorigenic: a conundrum. Clin Chim Acta. 2021;514:1–7.
73. Gao H, Liu C, Ren Q, Zhang L, Qin W, Wang H, et al. The novel SLRP family member Lumican suppresses pancreatic cancer cell growth. Pancreas. 2023;52(1):e29–36.
74. Brézillon S, Radwanska A, Zeltz C, Malkowski A, Ploton D, Bobichon H, et al. Lumican core protein inhibits melanoma cell migration via alterations of focal adhesion complexes. Cancer Lett. 2009;283(1):92–100.
75. Stasiak M, Boncela J, Perreau C, Karamanou K, Chatron-Colliet A, Proult I, et al. Lumican inhibits SNAIL-induced melanoma cell migration specifically by blocking MMP-14 activity. PLoS One. 2016;11(3):e0150226.

76. Brezillon S, Zeltz C, Schneider L, Terryn C, Vuillermoz B, Ramont L, et al. Lumican inhibits B16F1 melanoma cell lung metastasis. J Physiol Pharmacol. 2009;60(Suppl 4):15–22.
77. Coulson-Thomas VJ, Coulson-Thomas YM, Gesteira TF, Andrade de Paula CA, Carneiro CR, Ortiz V, et al. Lumican expression, localization and antitumor activity in prostate cancer. Exp Cell Res. 2013;319(7):967–81.
78. Pietraszek K, Brézillon S, Perreau C, Malicka-Błaszkiewicz M, Maquart FX, Wegrowski Y. Lumican – derived peptides inhibit melanoma cell growth and migration. PLoS One. 2013;8(10):e76232.
79. Chakravarti S, Magnuson T, Lass JH, Jepsen KJ, LaMantia C, Carroll H. Lumican regulates collagen fibril assembly: skin fragility and corneal opacity in the absence of lumican. J Cell Biol. 1998;141(5):1277–86.
80. Kao WW, Liu CY. Roles of lumican and keratocan on corneal transparency. Glycoconj J. 2002;19(4–5):275–85.
81. Bettin A, Reyes I, Reyes N. Gene expression profiling of prostate cancer-associated genes identifies fibromodulin as potential novel biomarker for prostate cancer. Int J Biol Markers. 2016;31(2):e153–62.
82. Mondal B, Patil V, Shwetha SD, Sravani K, Hegde AS, Arivazhagan A, et al. Integrative functional genomic analysis identifies epigenetically regulated fibromodulin as an essential gene for glioma cell migration. Oncogene. 2017;36(1):71–83.
83. Mayr C, Bund D, Schlee M, Moosmann A, Kofler DM, Hallek M, et al. Fibromodulin as a novel tumor-associated antigen (TAA) in chronic lymphocytic leukemia (CLL), which allows expansion of specific CD8+ autologous T lymphocytes. Blood. 2005;105(4):1566–73.
84. Levens E, Luo X, Ding L, Williams RS, Chegini N. Fibromodulin is expressed in leiomyoma and myometrium and regulated by gonadotropin-releasing hormone analogue therapy and TGF-beta through Smad and MAPK-mediated signalling. Mol Hum Reprod. 2005;11(7):489–94.
85. Khan FU, Owusu-Tieku NYG, Dai X, Liu K, Wu Y, Tsai HI, et al. Wnt/β-catenin pathway-regulated Fibromodulin expression is crucial for breast cancer metastasis and inhibited by aspirin. Front Pharmacol. 2019;10:1308.
86. Silva T, Gomes CP, Voigt DD, de Souza RB, Medeiros K, Cosentino NL, et al. Fibromodulin gene variants (FMOD) as potential biomarkers for prostate cancer and benign prostatic hyperplasia. Dis Markers. 2022;2022:5215247.
87. Gu HB, Sun L, Liu YX, Li N. Effect of silencing. Sichuan Da Xue Xue Bao Yi Xue Ban. 2020;51(2):165–70.
88. Oldberg A, Kalamajski S, Salnikov AV, Stuhr L, Mörgelin M, Reed RK, et al. Collagen-binding proteoglycan fibromodulin can determine stroma matrix structure and fluid balance in experimental carcinoma. Proc Natl Acad Sci USA. 2007;104(35):13966–71.
89. Zheng Z, Granado HS, Li C. Fibromodulin, a Multifunctional Matricellular Modulator. J Dent Res. 2023;102(2):125–34.
90. Zheng Z, Zhang X, Dang C, Beanes S, Chang GX, Chen Y, et al. Fibromodulin is essential for fetal-type Scarless cutaneous wound healing. Am J Pathol. 2016;186(11):2824–32.
91. Jian J, Zheng Z, Zhang K, Rackohn TM, Hsu C, Levin A, et al. Fibromodulin promoted in vitro and in vivo angiogenesis. Biochem Biophys Res Commun. 2013;436(3):530–5.
92. Zheng Z, James AW, Li C, Jiang W, Wang JZ, Chang GX, et al. Fibromodulin reduces scar formation in adult cutaneous wounds by eliciting a fetal-like phenotype. Signal Transduct Target Ther. 2017;2:17050.
93. Stoff A, Rivera AA, Mathis JM, Moore ST, Banerjee NS, Everts M, et al. Effect of adenoviral mediated overexpression of fibromodulin on human dermal fibroblasts and scar formation in full-thickness incisional wounds. J Mol Med (Berl). 2007;85(5):481–96.
94. Jazi MF, Biglari A, Mazloomzadeh S, Kingston P, Ramazani A, Bazzaz JT, et al. Recombinant fibromodulin has therapeutic effects on diabetic nephropathy by down-regulating transforming growth factor-β1 in streptozotocin-induced diabetic rat model. Iran J Basic Med Sci. 2016;19(3):265–71.
95. Delalande A, Gosselin MP, Suwalski A, Guilmain W, Leduc C, Berchel M, et al. Enhanced Achilles tendon healing by fibromodulin gene transfer. Nanomedicine. 2015;11(7):1735–44.

Critical Conclusive Overview: SLRPs as DAMPs

10

Polly Lama

Abstract

More than two decades have passed since the discovery of the first SLRP decorin, and to date eighteen members of this protein family have been identified, many of which are now recognised as potential biomarkers of extracellular matrix dysregulation. SLRPs plays significant role as an initiator, a link protein molecule, and an inhibitor of various types of signaling pathway in diverse range of pathophysiological disease processes involving a spectrum of biological actions, such as host defense, regulation of gene expression, cellular homeostasis, wound healing, inflammation, autoimmunity, oncogenesis, and collagen fibrillogenesis. SLRPs as a signaling molecule mediate interactions with various growth factors, cell receptors, matrix-degrading enzymes, and many other matrix components involved in cell growth, morphogenesis, cell death, and immunity. SLRPs and their fragmented peptides can alter signaling pathways thorough affinity-based interactions within cell and tissue, involving specific tyrosine kinase and Toll-like receptors, underscoring their critical roles as damage-associated molecular patterns (DAMPs). This chapter presents a conclusive overview on the multi-functional role of SLRPs highlighting their potential recognition as DAMP, by the host immune system across various tissue types. This presented perspective thus aims to rationalize the impact of SLRP function as a key mediator of inflammatory signaling process.

P. Lama (✉)
Department of Anatomy, Sikkim Manipal Institute of Medical Sciences, Sikkim Manipal University, Gangtok, Sikkim, India
e-mail: polly.lama@smims.smu.edu.in

P. Lama (ed.), *The Biological Role of Small Leucine-Rich Proteoglycans (SLRPs) in Matrix Homeostasis and Diseases*,
https://doi.org/10.1007/978-981-96-7890-7_10

Keywords

Small Leucine Rich Proteoglycans · DAMPs · Inflammation · Decorin · Lumican · Biglycan

10.1 SLRPs Act as DAMPs

Damage-associated molecular patterns (DAMPs) in the recent years have emerged as a tissue-sensitive novel mechanism that initiates and promotes inflammation by regulating innate and adaptive immune responses [1, 2]. DAMPs are expressed and released during tissue damage and cellular stress reactions, and many molecular mediators designated as DAMPs detect early signals in the stages of inflammation [3]. SLRPs can be classified as a subset of complex endogenous DAMPs, as their expressions are highly upregulated or downregulated in various diseases and are locally released in almost all inflammatory disorders associated with macrophage activation, such as rheumatoid arthritis, skin allergies, cardiovascular diseases, autoimmune diseases, osteoarthritis, degenerative disc disease, neurodegenerative disorders, local and systemic infections, tumors, and cancers of various organs and tissues [4], and such changes are not observed in healthy tissues. Many SLRPs promote inflammation via the activation of Toll-like receptors 2 and 4 [5, 6], and serum concentration level of decorin, biglycan, lumican, and fibromodulin increases in many types of cancers; hence, SLRPs are regarded as conventional biomarkers for the monitoring disease stages and activity, especially in osteoarthritis [7] and cancers [8].

In murine model of osteoarthritis, SLRPs has been shown to play a pivotal role in regulating inflammatory responses, and the release of SLRPs and their fragmented peptides has been found to correlate with disease activity in various pathological disorders emphasising their translational relevance and impact in both basic research and clinical applications, as discussed in the previous chapters.

SLRPs show multivalent binding capacity required for regulating normal functions, such as controlling collagen fibrillogenesis, and in addition they also form complexes with GAG side chains essential for maintaining the structural integrity of the tissues. The unique binding abilities of SLRPs in tissue disorders also enable them to act as adhesion site for pathogens and tissue toxins. SLRPs, such as decorin and biglycan, regulate growth factors like TGFs and BMPs facilitating the association of low-density lipoprotein and apolipoproteins to collagen, and this interactions contribute to the accumulation of these toxic mediators in atherosclerosis.

As described in the previous chapters SLRPs are involved in cell-matrix interactions in various tissues and organs, and also inhibit cell attachment in various types of cancers, osteoarthritis, and multiple types of tissue injury. Lumican, osteoadherin, and chondroadherin exhibit a high affinity for integrin and toll-like receptors, which further engages with matrix-degrading enzymes accelerating tissue degeneration. Biglycan regulates bone and muscle cell activity through its glycosaminoglycan side chains, thus playing a role in muscular dystrophies and bone formation.

Decorin, asporin, and nyctalopin, are cell membrane-bound SLRPs, that act by integrating tyrosine kinase cell receptors and the pericellular matrix molecules to alter cellular behaviors. SLRPs also act as cytokine reservoirs in the extracellular matrix, restricting molecular diffusion [9], which promotes fragmentation of the core protein by matrix-degrading enzymes and facilitates direct interaction with cell- surface receptors. Therefore, SLRPs influence cellular homeostasis to induce differentiation, apoptosis, proliferation, and migration through multiple signaling pathways *(already described in the previous chapters)*, altering the matrix, and thereby compromises the structural integrity of the tissues and organs.

10.2 SLRPs in Inflammatory Catabolism

Growing evidence indicates a significant role of SLRPs as a direct and indirect endogenous modulator of inflammation and innate immune responses. Biglycan, gets sequestered in the matrix by proteolytic cleavage and can exist as a soluble molecule in the injured tissues and are also secreted by activated macrophages. Similar to the action mediated by pathogen-associated molecular patterns (PAMPs) soluble biglycan acts as an endogenous ligand of Toll-like receptor 2 and 4 in macrophage, leading to activation and downstream signaling, via p38, p42/44, and NF-κB pathways, resulting in increased synthesis and secretion of pro-inflammatory cytokines. TNFα enhances recruitment of macrophages and neutrophils, which subsequently secrete additional biglycan, thereby creating a positive feedback loop that amplifies both autocrine and paracrine inflammatory signals. Biglycan null mice, show increased survival during sepsis, owing to the absence of TLR 2- and 4-dependent pathways. Clinically these null mice show reduced plasma levels of TNFα with reduced infiltration of mononuclear cells in the lungs, a major target organ affected by sepsis in mice. In noninfectious inflammatory renal diseases and osteoarthritis, overexpression of biglycan is associated with increased numbers of infiltrating mononuclear cells, and soluble biglycan activates inflammasome by the action of caspase-1 and IL-1β without any further need for additional costimulatory factors. Similarly, in inflammation of renal and lung tissue including injury, and lipopolysaccharide-induced sepsis, lack of biglycan is associates with active caspase-1 and mature IL-1β. Thus, biglycan, decorin, and lumican, upon their release from the matrix, act as an autonomous trigger of inflammatory response reaction (described in detail in Chaps. 3, 4, 5, 6, 7, and 8). In nonpathogenic inflammation, SLRPs engage with Toll-like receptors, which are actually pathogen-sensing receptors, SLRPs thus modulate the innate immune responses indirectly and can even influence TLR signaling by presenting PAMPs to the various receptor complex. This is demonstrated by the ability of lumican core protein that binds and presents lipopolysaccharide to CD14, thereby leading to the activation of TLR-4 pathway. Furthermore, SLRPs biglycan, decorin, fibromodulin enhance immune system activation by their ability to modulate TGF-β, a potent immunosuppressive cytokine. SLRPs biglycan and decorin can also modulate inflammation through macrophages and they can even reverse their repressive effect by inhibiting proliferation, apoptosis and enhancing the synthesis of proinflammatory cytokines, including nitric oxide synthase. Thus, the antiproliferative and antiapoptotic effects of SLRPs on macrophages are mediated through cell cycle inhibitors such as the cyclin-dependent kinases p27(Kip1) and p21(Waf1) which plays key role in cell cycle regulation, and additionally decorin has been shown to suppress the effects of the macrophage colony-stimulating factors.

SLRPs are also involved in the recruitment of immune cells to the site of injury directly by acting as ligands to cell surface receptors and indirectly by forming complexes with circulating signaling molecules and decorin stimulates the production of mononuclear cell-recruiting chemokine, sustaining the inflammatory phase. Many SLRPs, besides their interaction with Toll-like receptors, also act as a ligand for selectin, which is directly involved in the recruitment of natural killer cells [10]. SLRPs also binds and signal via the Fas ligand, which enhances the synthesis and secretion of proinflammatory cytokines and the recruitment of macrophages and neutrophils, involving the chemokines, thus establishing a gradient that regulates infiltration of immune cells. Furthermore, decorin, biglycan, fibromodulin, osteoadherin, and chondroadherin bind to C1q, resulting in activation of the classical complement pathway that leads to an enhanced inflammatory response [11].

Thus, SLRPs function as DAMPs and facilitate the presentation of PAMPs to the receptor complexes thorough interaction with various cytokines, chemokines, and complement factors to modulate innate and adaptive immune responses in infectious, noninfectious, and pathogen-mediated inflammatory conditions.

Although a convincing translational strategy targeting SLRPs is yet to be fully developed, emerging evidences hold promise. One such potential avenue involves integrating molecular imaging with hybrid diagnostic systems to acquire comprehensive information in a single examination. This is perhaps a future that can be sought, as the paradigm leveraging molecular imaging and integrated diagnostic approaches across various preclinical models of inflammation may pave the way for establishing the translational potential of SRLPs as a viable diagnostic biomarkers in clinical practice.

References

1. Chan JK, Roth J, Oppenheim JJ, Tracey KJ, Vogl T, Feldmann M, et al. Alarmins: awaiting a clinical response. J Clin Invest. 2012;122(8):2711–9.
2. Bianchi ME. DAMPs, PAMPs and alarmins: all we need to know about danger. J Leukoc Biol. 2007;81(1):1–5.
3. Schaefer L. Complexity of danger: the diverse nature of damage-associated molecular patterns. J Biol Chem. 2014;289(51):35237–45.
4. Roh JS, Sohn DH. Damage-associated molecular patterns in inflammatory diseases. Immune Netw. 2018;18(4):e27.
5. Schaefer L, Iozzo RV. Biological functions of the small leucine-rich proteoglycans: from genetics to signal transduction. J Biol Chem. 2008;283(31):21305–9.
6. Iozzo RV, Schaefer L. Proteoglycans in health and disease: novel regulatory signaling mechanisms evoked by the small leucine-rich proteoglycans. FEBS J. 2010;277(19):3864–75.

7. Flannery CR. Usurped SLRPs: novel arthritis biomarkers exposed by catabolism of small leucine- rich proteoglycans? Arthritis Res Ther. 2006;8(2):106.
8. Appunni S, Anand V, Khandelwal M, Gupta N, Rubens M, Sharma A. Small leucine rich proteoglycans (decorin, biglycan and lumican) in cancer. Clin Chim Acta. 2019;491:1–7.
9. Chen S, Birk DE. The regulatory roles of small leucine-rich proteoglycans in extracellular matrix assembly. FEBS J. 2013;280(10):2120–37.
10. Merline R, Schaefer RM, Schaefer L. The matricellular functions of small leucine-rich proteoglycans (SLRPs). J Cell Commun Signal. 2009;3(3–4):323–35.
11. Sjöberg AP, Manderson GA, Mörgelin M, Day AJ, Heinegård D, Blom AM. Short leucine-rich glycoproteins of the extracellular matrix display diverse patterns of complement interaction and activation. Mol Immunol. 2009;46(5):830–9.

MIX
Papier aus verantwortungsvollen Quellen
Paper from responsible sources
FSC® C105338

If you have any concerns about our products,
you can contact us on
ProductSafety@springernature.com

In case Publisher is established outside the EU,
the EU authorized representative is:
Springer Nature Customer Service Center GmbH
Europaplatz 3, 69115 Heidelberg, Germany

Printed by Libri Plureos GmbH
in Hamburg, Germany